THE XENO CHRONICLES

Two Years on the Frontier of Medicine

Inside Harvard's Transplant Research Lab

G. WAYNE MILLER

PublicAffairs
NEW YORK

Published in the United States by PublicAffairs™,
a member of the Perseus Books Group.

Book design and composition by Mark McGarry
Set in Mendoza

Library of Congress Cataloging-in-Publication data
Miller, G. Wayne.
The Xeno Chronicles: Two years on the frontier of medical science inside Harvard's transplant research lab/G. Wayne Miller.
p. cm.
Includes bibliographical references and index.
ISBN 1-58648-242-4
1. Xenografts. 2. Transplantation of organs, tissues, etc.
I. Title.
RD129.5.M55 2005   617.9'5—dc22   2005042567

FIRST EDITION
10 9 8 7 6 5 4 3 2 1

For the three most incredible children,
and the most incredible granddaughter:
Rachel, Katy, Calvin, and Isabella. I love you guys!!!

Contents

Introduction		ix
CHAPTER 1 :	Double-Knockout	3
CHAPTER 2 :	Hearts Stopping	25
CHAPTER 3 :	Waiting Rooms	65
CHAPTER 4 :	Risky Business	111
CHAPTER 5 :	The Value of a Life	164
CHAPTER 6 :	A Perilous Existence	188
	Epilogue	207
Acknowledgments		209
Notes		213
Index		221

INTRODUCTION

Some while ago, I set out to explore the frontiers of medical science.

I began with a visit to a molecular biologist who was unlocking the secrets of cellular aging, the kind of research that might lead to the age-old dream of a Fountain of Youth. From there, I traveled to two tissue engineers who aim to create organs and other body parts from technology that marries living cells to synthetic materials. These scientists are all preeminent in their fields, and their advances have potential for improving human life.

I was similarly impressed with the two leading stem cell scientists who welcomed me into their worlds. One experiments with embryonic stem cells, a politically controversial endeavor that someday, controversy notwithstanding, will play a significant role in regenerative medicine—technologies that repair or replace defective or damaged bodily parts and systems. The other scientist experiments with adult stem cells found in bone marrow and elsewhere in the body; he too is a key figure in bringing the promise of the frontier to the clinic, where people eventually will benefit.

Further out on the frontier, I connected with a scientist who has completed the conceptual designs for nanobots—microscopic robots that would act as white blood cells and other healing agents in a distant age. Someday, some nanoscientists believe, these tiny machines may allow us to live in good health for hundreds of years.

For those who die before such innovations actually become available, there is cryonics: deep-freezing bodies (or, for an economy price, heads only) with the intent of future resurrection. I visited the Scottsdale, Arizona, home of the world's largest cryonics company and gazed on the closed canisters in which dozens of corpses (and parts thereof) are stored in liquid nitrogen, temperature 196 degrees below zero. I came away unencouraged: technology can keep people frozen for decades, but the thawing end remains, shall we say, problematic. A writer friend put it in perspective when he noted that a steak kept for more than a few months on ice develops freezer burn that renders it useless.

And then I met Dr. David K.C. Cooper, a prominent researcher in xenotransplantation—the use of live animal tissues and organs to heal sick people.

Until I heard Cooper speak at a conference in the spring of 2002, my knowledge of xenotransplantation was mostly drawn from accounts I'd read about Baby Fae, the infant who received a baboon heart in a highly publicized operation in 1984. Baby Fae died soon after the transplant—and so, as far as I knew, had this seemingly quixotic concept. Perhaps a lone maverick somewhere was tinkering with the idea—a modern-day Dr. Frankenstein in his shadowy lab—but until meeting Cooper, I did not consider xenotransplantation more than a fringe science.

Xeno has a long and sometimes strange history, dating back at least to 1682, when a Russian doctor used pieces of dog bone to patch up the skull of a nobleman who'd been wounded in battle. The experiment was cut short when the Orthodox Church threatened the man with excommunication on the grounds that no true Christian could have part of a dog in his head. Rather than be forced from the church, the man had the bone removed.

By the early twentieth century, doctors were trying to put animal kidneys into people, and in the 1920s, a series of quacks transplanted animal testicles into aging men, claiming that the operations restored vitality. In the 1960s, legitimate surgeons experimented with implanting baboon and chimp organs in people. Some enjoyed modest success, but the immunological factors involved in cross-species transplants remained largely mysterious, and for a number of years, research was limited.

The rise of contemporary xenotransplantation research began with the introduction of cyclosporine, the drug that suppresses the body's rejection of foreign tissue. Introduced in the early 1980s, cyclosporine made conventional transplantation—human organ into human—a life-saving option for thousands of people who would have otherwise died. As a result, the demand for human organs began to rise and soon outstripped supply. Large pharmaceutical firms, including Sandoz, which sold cyclosporine, sought ways to close the gap. Animals were a logical place to look.

Cooper once worked as a heart transplant surgeon in South Africa with the great pioneer Dr. Christiaan N. Barnard, who transplanted the first human heart in 1967. Cooper moved on to Oklahoma and eventually to Boston, where he was a principal scientist in the Transplantation Biology Research Center, a division of Massachusetts General Hospital and an affiliate of Harvard Medical School. Cooper's work involved transplanting pig hearts into baboons. The goal was to perfect the science so that people would benefit.

At the spring 2002 conference, Cooper explained that xeno could help alleviate the chronic shortage of human organs. Each year thousands never reach the top of the transplant waiting lists—and thousands more are too sick to make it onto the lists at all. With stem cell and tissue engineering technologies still years from producing a whole organ such as the heart, kidney, or liver, xeno makes practical sense. If you could somehow manipulate the human immune system to accept an animal organ, all you'd have to do is order one from the biotech farm.

Xenotransplantation raises intriguing issues. What would a person with a pig heart think about it beating inside their chest? What would such a person be—part animal, part human, like a creature from Greek mythology—or something else entirely? What of the threat of disease being transmitted from animal to human—of some previously unknown virus that could spread into the general population before it was recognized? AIDS and *The Hot Zone* came to mind. What about animal rights activists who oppose any "exploitation" of animals, let alone farms where they would be raised for their organs? What about the ethics of patients who might wish to use xenotransplantation to extend life past the limits imposed by "God's plan"? What other scientists were behind this rarefied branch of medical science?

Harvard backing or not, how realistic was xenotransplantation?

Cooper invited me to visit his lab, where I met his boss, Dr. David H. Sachs, in June 2002.

Sachs rarely grants interviews, and outside of medicine his name is virtually unknown. But within his field, transplantation immunology, Sachs is a near-legend. He and the scientists in his lab are responsible for numerous advances in conventional transplantation (human-to-human organs and tissues), and he arguably is the leading figure in xeno, which has been his professional passion for three decades.

The day we met, early in June 2002, Sachs and I discussed what I wanted to do: spend time with him so that I could write a firsthand narrative of his quest to bring xenotransplantation to the clinic. Although a handful of journalists have visited centers where nonhuman primates are the subjects of experiments, none, to my knowledge, was ever invited to stay—certainly not for the two years I was there. But Sachs felt his work was vital and deserved a larger audience than fellow scientists and readers of the scientific literature.

And so, after consideration, Sachs agreed to give me unlimited access to his lab, his animal facility, his people, and himself. His only prohibition was against photographing the baboons. He feared that animal rights activists would obtain copies and alter them for use in their cause.

As the weeks turned to months and I all but joined Sachs's staff, I discovered dimensions to the story that I hadn't anticipated.

The money angle was one. As in other cutting-edge research, big business plays a significant role in funding xeno experimentation, which is expensive, in part because large animals are involved. Continued funding depends on results, which pressured Sachs and the small biotech firm he worked with, Immerge BioTherapeutics. And "results" didn't mean just scientific advances. As one corporate executive who controlled some of the resources that Sachs received told me, a for-profit company at some point has to bring a product to market. It has to make money, no matter how noble the cause. This is capitalism, and shareholders must be satisfied.

I also learned that after some three decades of work in xenotransplantation, Sachs's best hope for a breakthrough had come down to a twenty-one-pound pig.

The pig had been cloned from a cell that had been engi-neered to remove or "knock out" both copies of a gene that produces a sugar molecule that sets off a violent rejection when organs are transplanted into another species. Sachs believed that the new pig and others like it that were being produced would not trigger this destruction. He believed that the animal might also provide protection against more delayed forms of rejection—rejection that sets in days or months later—that can also destroy a transplanted organ.

This pig excited Sachs—success might head xeno toward clinical trials, the last step before animal organs would be available to anyone in need of a transplant. "If everything went as well as could possibly be expected," Sachs told me shortly before the first experiments with the new pigs began, clinical trials "would probably be another year and a half. That's the best-case scenario."

And the worst?

"That nothing we do with this new pig is any different from what it was with the non-knockout," Sachs said. "And we find that there are other antigens, other antibodies, other mechanisms that are just as powerful—and this anti-body that we have thought was the most important barrier was really only one of several barriers."

That would mean more research, which would mean more time and money. But Sachs was by nature an optimist, regardless of the odds.

*

And then there was David Sachs the person—a man in his early sixties who remained filled with a childlike sense of wonder at nature, especially the human body, with its immune system and brain and musculature and all its other features. In this sense, he is like the great eighteenth-century British surgeon-scientist John Hunter, whose intellectual curiosity knew no bounds.

"I marvel every day," Sachs said. "I wonder at the universe. I wonder at all these things."

Blessed with professional success and a loving family, Sachs had few regrets. Chief among them was the knowledge that he would not live to experience the wonders of ages to come.

"I don't resent the fact that there's so much we don't know," he said. "In fact, I'm awed by it. What I resent is the fact that I won't be around to see the things understood that I want to understand and to see the answers. It makes me sad. And the older I get, the sadder it makes me. I love to think what will happen—all of what we're doing, transplantation, will disappear someday because we'll have ways of regenerating organs. There will be so many advances, but it's going to be so long in the future. But it will happen."

And so, he believed, would xenotransplantation. I present *The Xeno Chronicles* as an account of scientists on the far frontier at a critical moment for their field.

G. Wayne Miller
Pascoag, R.I.

THE XENO CHRONICLES

CHAPTER 1

Double-Knockout

A Twenty-One-Pound Pig

A cold day was dawning when Dr. David H. Sachs left his home and headed to his Boston laboratory, a few miles distant. He was praying that experimental animal 15502—a cloned, genetically engineered pig—had arrived safely overnight from its birthplace in Missouri.

It was Friday, February 7, 2003.

Ordinarily a calm and measured man, Sachs had fretted for weeks over this young animal, whose unusual DNA might help save untold thousands of human lives. He worried about the weather, so frigid that Boston Harbor had iced over and pipes in the animal facility had frozen, fortunately without harm to the stock. He worried that the pig would become sick, so he decided against transporting it by truck, for a winter storm could prove disastrous. Then he worried about flying it up. What type of aircraft should they use? Commercial? Charter? Which airport in the congested metropolitan area would be safest?

"Use your best judgment," Sachs had told the staff veterinarian assigned to bring the pig north. "Just don't lose this pig!"

An immunologist who trained in surgery, Sachs had distinguished himself in the field of conventional transplantation with human organs. His lab, the Transplantation Biology Research Center, was a part of Massachusetts General Hospital, where he was on staff. He was a professor at the Harvard Medical School. He belonged to the National Academy of Science's Institute of Medicine. He was fluent in four languages. He had written or cowritten more than seven hundred professional papers. Science came as naturally to him as breathing.

One achievement, however, still eluded Sachs. For more than thirty years, he had tried to find a way to get the diseased human body to accept parts from healthy animals.

Xenotransplantation—cross-species transplantation—has the potential to save thousands of people who die every year because of the chronic shortage of human organs. Children born with defects and older people with all manner of ailments would benefit. Sachs himself was not motivated by money, but xeno could become a multibillion-dollar business. You couldn't buy or sell a human organ, at least not in America and most countries of the world. But there were no laws in the United States against commerce in animal parts, although animal rights activists and others believed there should be.

Many scientists over many years had tried to achieve what Sachs sought.

So far, the idea remained a dream.

A man of average height, Sachs was on a diet but still carried a few too many pounds, a fact he jokingly acknowledged when describing himself as "chubby." With his full, graying hair and his jolly face, he evoked the image of a teddy bear— especially when he laughed, which was often. Sachs favored button-down shirt, tie, khaki slacks, a rumpled suit jacket, and wingtip shoes. Unless you looked closely, you would not notice that his right foot was larger than his left, a remnant of polio, the scourge of the 1940s, which he contracted when he was four and a half. Sachs had spent weeks of his early childhood at Manhattan's Hospital for the Ruptured and Crippled, an institution on 42nd Street and Lexington whose very name evoked suffering. The doctors said he might never walk again. But he did, perfectly normally.

"It just never seemed possible to me that I wouldn't," Sachs said. "It just seemed to me that I had to get over this problem. I've never had a defeatist attitude toward anything. I always feel that it's just a matter of being able to figure it out, make it work. That's my attitude toward everything."

But Sachs could not stop the clock. He was past sixty now and increasingly conscious of his mortality.

"Only recently have I started to realize how finite my lifespan is," he said. "Of course I've known that since I could think, but as you get older you realize that nobody lives past one hundred and I'll be lucky if I get over eighty. So I don't have a hell of a lot of time left."

In his darker moments, Sachs worried that he would never achieve his grand ambition. Experimental animal 15502—a creature small enough to fit into a baby stroller—might well represent his last chance.

Sachs parked his Saturn, then cleared the guards with their walkie-talkies and closed-circuit TVs in the lobby. He needed a card key to operate the elevator and to open the outer door to his lab, owned by Mass General Hospital, on a floor upstairs. Another lock secured the administrative suite, with yet one more protecting his inner office.

Once inside, the atmosphere was delightful. This was a bright, open corner space with a commanding view of Boston Harbor, and Sachs had decorated it with photographs of colleagues, mentors, friends, and his wife, Kristina, and their four children. Except for the books and professional journals, one of the only clues into the nature of Sachs's work was a pink stuffed pig.

It was nearing eight o'clock.

Sachs checked his e-mail messages, most of them concerning the status of experiments and grants. He put on a lab coat, left his office, and passed through the main conference room, which featured a photograph of a Little League team that he had sponsored and a bulletin board for laboratory business.

Alongside the ordinary notices and schedules was a clip of one of the few stories about Sachs to appear in the mainstream press. It was a piece from the August 11, 2001, *San Francisco Chronicle,* and it concerned a precursor animal to 15502. "This little piggy may be what the doctor ordered," the headline read. The clip included a photo of Sachs, posed behind a beaker. The writer listed some of the reasons the miniature pig Sachs used was the current focus of xeno research: its organs are similar in size and function to a human's, and nearly 100 million pigs are slaughtered for food every year in the United States alone, with little protest from anyone. And unlike chimpanzees, who were immunologically closer to people and resembled people more than any other animal, pigs were, well, pigs.

The *Chronicle* story was straightforward and informative, but a printout of an Internet page that someone had posted suggested xeno could prompt a quirky humor. "The U.S. is critically low on organ donations. What is the nation's medical community doing to address the shortage?" the page read. One answer offered was, "Removing David Crosby's new liver and giving it to a more deserving person." Another was, "Allowing recipient's body to reject maximum of two hearts; after that, no more favors for Mr. Picky." A third was, "Experiment with tofu-based organ substitutes." Xeno humor wasn't confined to this bulletin board. An unsigned

editorial comment in a recent issue of *Xenotransplantation*—
a professional journal founded by Sachs and now edited by a
senior member of his staff—noted that some surgeons were
considering face transplants. "This is unlikely to be an area
in which the xenotransplanters can become involved," the
editorialist wrote.

Sachs left the conference room and went to a separate
facility, secured by another electrically locked door. He
opened it with his card key and stepped into a windowless
domain constructed of cinderblocks painted a glossy insti-
tutional beige and lit with fluorescent bulbs. This was the
animal area, where experiments were conducted and where
he gathered his scientists every Friday at eight for large-animal rounds.

At any given time, Sachs's staff included nearly ninety
scientists, technicians, assistants, administrative aides, and
secretaries, whose salaries were funded by government
grants, university fellowships, or money from corporate
sources. Many were research fellows—young men and
women who stayed a year or two before moving on in their
careers. Competition for these fellowships was intense. Over
the years, Sachs had attracted scientists from all over the
world.

Some two dozen men and women, most wearing white
coats, awaited Sachs on that February 7. They lined a wall of
the animal area's central corridor, which separated the
baboon room and the pig rooms from the operating suites.
A strict, if unwritten, hierarchy was observed: nonscientists

stood at the ends of the line, with the middle held by two of Sachs's senior staff researchers. One was Dr. David K.C. Cooper, sixty-two, a tall, refined British surgeon (the K.C. stood for Kempton Cartwright). The other was Dr. Kazuhiko Yamada, forty-three, a Japanese surgeon who so impressed Sachs with his surgical prowess and innovative ideas during his 1990s fellowship that Sachs brought him on staff. Both doctors conducted research on allotransplantation (same-species transplants), but they, like their boss, had a special passion for xeno.

Sachs took his place in the precise center of the line, bid everyone good morning, and large-animal rounds were under way.

One by one, scientists stepped to a chalkboard in the corridor that held dozens of paper cutouts shaped like pigs. Each cutout had a number that corresponded to an animal. The scientists described the status of each experiment and Sachs, who had no notes but held the most minute details in his head, asked questions and made suggestions. Although he himself no longer conducted experiments, Sachs supervised all that took place in his lab. He was not overbearing or arrogant, just frightfully smart. Only when one of his people was clearly headed in the wrong direction, which was infrequent, would he overrule.

Most experiments involved transplanting organs from one pig to another, a model that mimicked conventional human transplantation. In their efforts to make transplants simpler and safer, Sachs's researchers experimented with drugs, radiation, bone marrow, and the thymus, a small gland that plays a large role in the immune system. The Holy Grail of all their experiments was tolerance—eliminating the lifelong need for immunosupressive drugs that a recipient must take to prevent rejection. These drugs can cause cancer and many other side effects, some potentially deadly and some cosmetically unappealing, and they leave a recipient at increased risk of infection. "A cold that you or I would get over in a day or two can be life-threatening," Sachs said. Weight gain and hair growth didn't kill, but for women especially, they could be depressing.

With tolerance, Sachs had enjoyed a measure of success. Working with colleagues at Mass General and building on a long body of research by other doctors, he had devised a way to manipulate the immune system of certain recipients to accept a transplanted organ without having to remain on drugs. The method involved bone marrow transplantation and was one of many attempts to achieve tolerance. After years of research with pigs, Sachs helped move tolerance from the lab to the clinic—where a few people had already benefited, including a woman who was still alive and well and immunosupressive-free three years after a kidney transplant.

The woman, Janet McCourt, a Massachusetts resident, was the first patient to try the tolerance protocol devised by Sachs and his colleagues.

She had been on dialysis and hated its constraints—being married to a machine was not her idea of living—and she was willing to try anything to get off, including a treatment attempted only on laboratory animals. "They told me the odds were unknown, because this had never been tried in a human," the middle-age McCourt told a Mass General publication. "But when we found that my sister was an excellent match and that she was willing to be the donor, I decided to go ahead and risk it. If I couldn't have the kind of active life I wanted, if I couldn't play with my grandchildren, I simply didn't want to live."

More work remained to be done so that more people could benefit—the best results so far were with closely related family members. But success with McCourt and a handful of others was nonetheless a breakthrough for Sachs and his colleagues. It was big news, not only in the scientific literature but in the mainstream press. It was the sort of work a Nobel committee might pay attention to.

But the excitement at this morning's large-animal rounds was not for bone marrow–induced tolerance. It was for the overnight arrival of a certain twenty-one-pound pig.

"Goldie is here," veterinarian Mike Duggan announced.

Sachs, of course, was aware of xeno's long, colorful past. He knew that there had been renewed enthusiasm for xeno-transplantation in the 1990s, when scientists made major advances and large corporations invested heavily in the field, only to be disappointed when xeno failed to reach the clinic—and return anything on their investments. He knew that some people, animal rights activists especially, had ethical objections. He knew of scientists and laypeople who foresaw dire public health consequences if some virus or other pathogen spread from pig organs into the human population.

Sachs knew too of other emerging technologies in regenerative medicine: stem cell science, tissue engineering, and artificial organs. He knew all this and still believed that xeno would play a major role in health care if it could be perfected.

"I see xeno as a possible answer to the organ shortage in the short term," he said. "Adult stem cells and tissue engineering have great appeal because they can potentially provide organs that are composed of 'self' tissues, thus avoiding the immune response. Fetal or embryonic stem cells are creating a lot of hype, but the tissues and organs derived from these sources would be just as foreign as allotransplants.

"For simple tissues—like islets or skin—stem cells and tissue engineering may provide solutions in the near future. However, for organs I think we are many, many years away from a solution. There is too much we don't know about the intra- and intercellular signaling required to make a complex organ to expect rapid development of the technology. Artificial organs have the problem of a power supply, and until we can harness nuclear energy in a practical form, organs like the heart will require too large a battery pack to be practical. Also, people do not want to be dependent on an external source of energy that could be interrupted, e.g., by blackouts."

Xenotransplantation had its supporters from outside the field, notably embryonic stem cell researcher Dr. Robert P. Lanza, chief scientist for Advanced Cell Technology, the small biotech firm in Massachusetts that claimed to have cloned the first human embryo. Lanza had demonstrated in the mouse that stem cells could be used to repair muscle damaged in heart attacks; someday soon, he predicted, the same technology could be used for people. But it would be many years, he also believed, before stem cell science could create an entire replacement heart—or kidney, lung, or liver, for that matter.

"To come up with an entire heart for someone who, say, has congestive heart failure—right now other than normal transplantation the only option is xenotransplantation," Lanza said. "For the next twenty years, there is going to be a window of opportunity for xenotransplantation."

Goldie

Starting in 1973, when he was at the National Institutes of Health, Sachs had bred a line of miniature swine for use in his many transplant experiments. By now, he had bred more than 10,000 pigs, and his current colony numbered about 450, but he had never named one. It served no good purpose, he reasoned, for anyone to become emotionally attached to a creature unlikely to see old age.

But 15502 was unlike any of Sachs's other pigs.

After years of trying, scientists at Immerge BioTherapeutics, a Massachusetts biotechnology firm with which Sachs collaborated, had recently succeeded in knocking out both copies of the gene that produces a sugar molecule found on the surface of ordinary pig cells. These sugar molecules are harmless to the pig. But when a pig organ was transplanted into a baboon (or a person), the recipient's immune system recognized them as the calling cards of an invader, a discovery made by Cooper and colleagues in Oklahoma in 1991. Within minutes, white blood cells attacked and destroyed the organ, leaving it a dark, useless mess. Hyperacute rejection was the scientific term for this vengeance, which evolution created as a life saver: the same sugar found on pig cells is also found on the surface of certain parasites, viruses, and bacteria, some of which are fatal to the human. Evolution had not anticipated transplantation.

Sachs also hoped that fewer drugs would be needed with this pig's organs. That would be another step toward tolerance—the Holy Grail.

Sachs and Immerge also collaborated with the National Swine Research and Resource Center at the University of Missouri's College of Agriculture, Food, and Natural Resources. It was scientists there who cloned animal 15502 from a cell lacking both copies of the sugar gene that Immerge BioTherapeutics had supplied. The Missouri scientists wanted to name the pig, born a week before Thanksgiving, and the name they chose for the so-called double-knockout pig was Goldie. The name captured the medical promise of the piglet and its commercial prospects as well.

Immerge BioTherapeutics aimed to get a decent share of that business.

Duggan recounted his journey to fetch Goldie, and it was good that Sachs was learning the details after the fact.

The flight down, in a chartered twin-engine plane that had left Worcester airport the morning before, Duggan said, was uneventful—at first. But just after refueling in Ohio, the airplane's heater blew and the temperature in the cabin dropped below zero. "At that point I was more concerned about my well-being than any pig!" Duggan said. Three hours later the plane touched down safely in Missouri, where mechanics fixed the heater and a veterinarian drove up with Goldie.

Hand-fed since birth and pampered perhaps more than any other experimental pig, Goldie had never before left her home.

Now she was inside a dog crate, flipping out.

"She had always been in a single room by herself with a lot of human contact," Duggan said, "and now she was taken from that environment, put into a crate, put into the back of a truck, driven to an airport. Then she was being taken out of *that* environment into the strange environment of a plane—loud engines, the whole issue of takeoff, everything. All of that was very stressful." Duggan feared that on the trip north, the stress would induce shipping fever—a pneumonia-like affliction, well-known to horse breeders and veterinarians, that can kill. Duggan considered sedating Goldie, but talking to her and offering her a bottle calmed her down, and no sedative, which carried its own risks, was needed. "The same way as if I'm flying with my own child," the veterinarian explained. Three-month-old Goldie also found comfort in her ball and teddy bear. Soon she was asleep.

Skirting a storm, the plane returned to Worcester at about 9:00 P.M. Goldie was transferred to a van and driven to Sachs's lab, where she was placed—with her bottle, ball, and teddy bear—in a cage lined with lambswool. The cage provided Goldie with a constant flow of filtered air that kept the dust and germs away. Duggan settled the pig in and hand-fed her a dinner of fruit. It was nearing midnight when he dimmed the lights. "Goodnight, I'm off to bed," Duggan said. "See you in the morning." Goldie passed a restful night and was happy and playful at breakfast that morning.

Duggan told the group in the corridor, "She does interact well with people." It would be fine to pet her, he said—with gloves.

"I don't think you need to pet her," Sachs said.

"It should be minimized," Duggan agreed. "It shouldn't be like a circus back there."

Animal rounds ended with a review of the current xeno experiments, which involved transplanting organs that had the sugar gene from pigs into baboons, whose immune systems are similar to a human's. Although Goldie and others like her that would be produced seemed to be the best chance of solving the xeno puzzle, other approaches using other protocols and another type of genetically modified pig had been tried before double-knockouts existed, at Sachs's center and elsewhere. The best success was a transgenic pig heart that Cooper transplanted into a baboon that beat for 139 days before being rejected. Sachs had obtained these particular pigs, whose genome contained human DNA that scientists had microinjected into the nuclei of fertilized eggs, from Imutran, a British xeno firm that had developed them in the 1990s.

But neither Cooper nor anyone else had been able to get another pig heart to last anywhere close to 139 days, and he, like his boss, believed the best chance was with double-knockout pigs.

Sachs and the senior members of his staff pulled on shoe covers to protect the animals from germs that shoes can carry and passed through an electrically operated door into the larger of the lab's two pig chambers. The room had dozens of cages, some plastic walled, some with steel bars. Most were occupied: one pig per cage. Some pigs still wore bandages from recent operations, and most had intravenous lines for administering medications and taking blood samples. The room was clean and bright with only the faintest trace of odor. Sachs treated his animals with compassionate care, for they were his most valuable tools. Inspections by federal and hospital agents, who often showed up unannounced, rarely disclosed infractions.

Goldie stood in her special cage, her ball and teddy bear at her feet. "Should be a different color or something, don't you think?" said a scientist. But she was a pretty shade of pink like the other pigs, and she was uncommonly cute. She bore an eerie resemblance to Babe.

Sachs put on rubber gloves and reached into the cage. Goldie came to him without hesitation.

Sachs patted the animal, felt her ears, tickled her snout. Goldie snorted agreeably but Sachs said nothing. He was listening to her breathing to confirm that her lungs were clear—that she was healthy.

"A lot of hopes are riding on this pig," Sachs said.

They were—not only for him and for the future of xenotransplantation but for a $25 billion company based in Switzerland. Novartis, the seventh largest drug firm in the world, was the principal investor in Immerge BioTherapeutics, a significant source of funding for Sachs's lab. Novartis (known then as Sandoz) had spent untold millions of dollars in the 1990s on Imutran, the English xeno firm. But Novartis closed Imutran in 2000 when its scientists failed to get pig organs to survive long enough in baboons to justify clinical trials. Sachs sometimes wondered how long Novartis would continue to support Immerge without a breakthrough.

Novartis's investment in Immerge BioTherapeutics was $10 million a year for three years, with an option to renew.

This was February of year 3.

A magnet held Goldie's paper cutout to the board at large-animal rounds the following Friday, Valentine's Day, but it would be a fleeting presence. This was the last such meeting before, as Duggan put it, "Goldie gives it up for science."

On Wednesday, February 19, the animal's heart, kidneys, thymus, and bone marrow would be transplanted into four separate baboons. As far as anyone in Boston knew, nothing like this had been attempted before—five animals, four operating tables, six surgeons, six technicians and assistants, one operating room supervisor, all on one day. Just getting the organs moved into their proper places would be an accomplishment. Sachs saw no value in a rehearsal—these were accomplished surgeons, after all—but he, Cooper, Yamada, and the staff had spent hours devising a written plan.

"Only people who are really essential go into the OR that day," Sachs said. "Don't come in unless you're asked."

Although Sachs didn't know whether anyone else had yet tried using a double-knockout pig in an experimental organ transplant, he knew that other such pigs existed. The Blacksburg, Virginia–based PPL Therapeutics, a subsidiary of the Scottish firm that commercialized the technology that created Dolly the sheep, the first mammal ever cloned, had competed in a highly publicized race with Immerge to create them. PPL won that race in the summer of 2002.

PPL also revealed that it was collaborating with the University of Pittsburgh Medical Center's renowned Thomas E. Starzl Transplantation Institution, overseen by the legendary surgeon himself, the man who had performed the world's first successful person-to-person liver transplant almost three decades earlier. Starzl had retired from the operating room but kept a hand in pushing the frontiers of medical science.

Several researchers around the world specialized in xeno-transplantation, but few had major corporate support or the resources of Harvard or the University of Pittsburgh. With Imutran gone, the Pittsburgh group was now Immerge's— and Sachs's—most formidable rival. Assuming xeno could be perfected, the group that brought xeno to the clinic first would claim not only scientific accolades but also a good share of a market that a Salomon Brothers study had predicted would reach $6 billion by the year 2010. The estimate did not seem unreasonable. No one could state what a working pig organ would cost, but with so many desperate patients and with waiting lists for all organs growing, the seller could all but command his price.

Starzl had a long-standing interest in the field. In 1964, three years before his historic first liver transplant, he tried six times to put baboon kidneys into people. All six died, but Starzl's interest in xeno persisted. As recently as 1992 and 1993, he and his team had attempted to transplant a baboon liver into a person. One patient lived seventy days but the other just twenty-six, and Starzl did not make another attempt. He did not lose his interest in xeno, however, although his failures had tempered him. "Our whole investigative team is working on almost nothing except xenotransplantation," Starzl told *The Scientist* in August 1995. "So, perhaps we have a strong insight about how tough this really is. We're not going to open up again until we have something that we think is a very fundamental improvement. Even though we've had permission to go forward, we're not nuts." Starzl knew the field needed a bold innovation before it could advance much further.

Word that Starzl was collaborating with PPL suggested that he was ready to try again. It suggested that the double-knockouts were what he'd been waiting for.

But no one in Boston knew exactly what Starzl and his scientists were up to, for they had published and publicly said little of substance since announcing the birth of the prize litter. Sachs was curious, of course, and gentlemanly though he might be, he was a player. In a world where academic researchers competed for corporate, philanthropic, government, and university dollars, and there were never enough to go around, it was impossible to succeed without being one.

At rounds on the Friday before Goldie's last day, Sachs asked, not for the first time, if anything had come down the grapevine regarding what Starzl was up to. One research fellow said he heard that the Starzl group had conducted some sort of xeno experiments with their double-knockout pigs, but the details were sketchy. Someone else heard that Starzl's first xenotransplants would be in late February or March. Someone joked that perhaps Starzl was planning on transplanting a liver into himself.

And maybe Starzl did have something dramatic like that in mind. He was one of the last of the old-time mavericks, a surgeon like Barnard and the open-heart pioneer Dr. C. Walton Lillehei, who came of age in the 1950s—when hospital ethics committees were nonexistent, patients rarely sued, and nosy reporters never called. When, in other words, you could get away with just about anything. Maybe Starzl intended to bypass baboons or monkeys and transplant his double-knockout organs directly into humans. Unlike Sachs, he had an officially sanctioned option on this score too, one that would not endanger a healthy or sick person's life: the University of Pittsburgh Medical Center had established a twelve-member Committee for Oversight of Research Involving the Dead to govern experiments on brain-dead people who were being maintained by machines. Other centers, including Temple University School of Medicine and M.D. Anderson Cancer Center in Houston, have similar committees governing similar research.

"It sounds macabre to manipulate a corpse," Arthur Caplan, director of the Center for Bioethics at the University of Pennsylvania, told a reporter when word of the committee leaked out. "But I tell you, it's worse to make someone a corpse." Caplan had a point.

The committee had already approved testing a kind of artificial lung on one of these zombies, a middle-age woman who was declared brain-dead after a car accident. When her organs were deemed unsuitable for transplant because they tested positive for hepatitis, Dr. Brack Hattler, who developed the artificial lung, asked the family for permission to test it. The family agreed—if they could watch. Hattler let them. It was after midnight when the doctor and his assistants administered a local anesthetic, an unnecessary but kind touch, to her leg and began threading a catheter-like device through a vessel to her heart. For five hours, working as designed, Hattler's machine gave out oxygen and drew back carbon dioxide. The team took blood samples. When they were done, the woman's ventilator was shut off, and she died, for good, a short while later.

The Hattler experiment reached the newspapers, but the Boston group could only conjecture what the close-lipped Starzl and his scientists were up to with PPL's double-knockout pigs.

"I imagine he wants to make some big splash," Sachs said. Starzl enjoyed making headlines.

Another concern was to prevent next Wednesday's events from being publicized. Sachs did not want word leaking out to reporters or animal rights activists, who, he maintained, could not fit the care he gave his research animals into the activist philosophy that animals should enjoy the same basic rights as people. Regardless of the cause—the betterment of people—animals did not deserve to be the subjects of experimentation, the activists believed.

Sachs ordered that inquiring calls be referred to him.

"When the time comes," he said, "I want the news to be correct."

But as it happened, reporters and animal rights activists would be among the least of the scientist's concerns.

CHAPTER 2

Hearts Stopping

Big Day

Goldie ate her last supper, a treat of dog food and raisins, early on the evening of Tuesday, February 18.

When attendant Shannon Moran arrived at seven o'clock the next morning, the pig was rooting around for breakfast, but Moran, following the standard preoperative orders, could not oblige. She administered a sedative and soon Goldie settled down on the softness of her lambswool. Moran swaddled the groggy pig in a sheet and hefted her up.

"All right, let's go, baby," she said.

Moran carried the pig into the hall, past a donor-awareness bumper sticker that read, "Don't Take Your Organs to Heaven; Heaven Knows We Need Them Here." Moran entered a small washroom between the two operating rooms and lowered the pig onto a stainless steel table, where technicians Crystal Dugan and Meaghan Sheils sheared Goldie's abdomen with well-worn barber clippers. At one point,

Goldie stirred. "Almost done! Go back to sleep!" Sheils said. Sheils had an uncommon touch with animals, whether named or numbered.

Goldie's hair was vacuumed away and the pig was transferred to the operating room—to a veterinarian's operating table, which is bowed like a shallow trough to better cradle an animal. After tying the pig's limbs to the corners of the table, Dugan and Shiels attached pulse and blood pressure monitors, started an intravenous line, and connected Goldie to the anesthesia machine. The bellows began their mechanized breathing.

It was 7:40 A.M., time to get under way.

On schedule, as he usually was, Dr. Kazuhiko Yamada— Kaz, as most called him—walked into the operating room. He would harvest and transplant most of Goldie's organs involved in the double-knockout experiments. Cooper would handle the heart.

"Big day!" Kaz said. Kaz spoke English fluently, albeit with an accent. His written English was impeccable.

The surgeon scrubbed Goldie's hairless belly with alcohol and two types of sterilizing surgical soap, then draped the pig; except for a glimpse of curly tail and two hind hooves, it could have been a child about to undergo a tonsillectomy beneath the sheet. Kaz photographed his chief assistant, Yosuke Hisashi, standing by Goldie, and then Hisashi photographed Kaz. Hisashi was a colleague from Japan whom Kaz flew in for this one day.

"I had several times performed kidney and heart transplants on pigs with Yosuke in Japan and he assisted me nicely," Kaz explained. "In order to assist my surgery and catch up to my speed, surgeons must be trained under me for several cases even if the surgeon has ten years of experience. Thus I chose him rather than others because he had received significant amounts of training with me in Japan prior to this case."

Kaz's insistence on an outsider ruffled some feathers, since there was no shortage of surgeons in Sachs's lab. But Sachs abided Kaz. "This is an exceptional case," he said at one meeting to the critics, some of whom believed Kaz could do no wrong in the eyes of the boss.

Kaz put his camera away, and he and Hisashi pulled on fresh sterile gloves.

It was nearing eight o'clock.

Across the central corridor—behind yet another locked, windowless door—things were busy in the baboon room.

This was a smaller area than the pig quarters, for, with space tight, the baboon population never exceeded sixteen. The baboons, like the pigs, each lived alone in a cage—but these cages were padlocked, and humans approached with caution. Baboons were strong and fast and some were threatening: Make eye contact and they might hiss and snarl and try to grab you through the cage bars. They were ugly creatures with angry faces and brown or gray fur, utterly lacking the charm of a chimp or the nobility of a gorilla. No moviemaker ever told the story of a cuddly baboon. No pig room featured a pig bite kit.

Most of the baboons were not headed for surgery today, and an attendant was feeding them a breakfast of oranges, apples, bananas, and sweet potatoes. A videotape played on a TV monitor. The baboon library included *Babe,* ironically enough, but this morning's presentation was *People of the Forest: The Chimps of Gombe,* based on the work of naturalist Jane Goodall, who often criticized the use of primates for biomedical research. It was difficult to determine if the animals were watching. Too much was going on that demanded their attention.

Except for the handful of baboons that enjoyed privilege as blood donors, these animals had no names. B214 watched quietly from a perch on the side of its cage as research fellow Dr. Kenji Kawaki, Cooper's chief assistant today, drew a blood sample from the intravenous line that ran inside a tether that was secured to the animal with a nylon jacket. At thirty pounds, B214 was an average baboon, a male born at a Florida breeding center on June 5, 1998.

Back in the operating room, Kaz made his first incision into Goldie.

Since moving in 1991 to Massachusetts General Hospital from the National Institutes of Health in Bethesda, Maryland, Sachs had spent Wednesdays at the hospital's main campus, where he could exchange ideas with practicing doctors and see patients. Spending time with only computers and animals, he believed, could diminish a medical scientist's appreciation of the true cause.

"There's something about seeing a problem, hearing a problem, and trying to figure out the answer to that problem that occurs when a person's life is involved that you don't get from just listening to it or reading about it," Sachs said. There was an emotional element too. One regret in choosing research over the practice of surgery, he said, "was [losing] the daily contact with patients and the feeling that every day I did something worthwhile for somebody—which is a wonderful, instant gratification that you have being a doctor." Wednesdays made up for some of that.

Still, on this Wednesday, Sachs was sorely tempted to stay with the baboons and the golden pig. But, concerned that his presence might add to the tension, he decided his best course was to keep faith in his people—should anything requiring him arise, he was, after all, but a phone call away.

And so the day began for Sachs as usual: he arrived at 7:00 A.M. at the main campus and went to the transplant unit, where two dozen surgeons, nurses, nephrologists, hepatatologists, infectious-disease specialists, cardiologists, social workers, and others filled a small room near the nurses' station. This was weekly patient rounds, a meeting of the brain trust.

Here, too, an unspoken hierarchy was observed, and Sachs took his seat in the front row next to Dr. A. Benedict Cosimi, chief of the hospital's transplantation unit, who collaborated with Sachs on research involving kidney, pancreas, and liver transplantation, Cosimi's specialties. Presiding over rounds was Dr. Paul S. Russell, a tall, white-haired, seventy-eight-year-old gentleman who pioneered transplantation at Mass General with a kidney transplant in 1964. Sachs greatly respected Russell, one of his early mentors. Photos of the doctor were displayed in Sach's office, and his laboratory library was named after him. And Sachs was proud to be the first Paul S. Russell/Warner-Lambert Professor of Surgery and Immunology at Harvard Medical School.

Some of the cases presented at transplant rounds were patients bound for happy endings, including those who sooner or later would be successfully transplanted and those who had already received their organ and would be going home, albeit with boxes of drugs they would have to take for the rest of their lives. Whenever a patient was discharged, Sachs and his colleagues celebrated the wonder of transplantation—and were reminded of the urgency of perfecting tolerance and xeno.

But other patients were desperately ill—children, women, and men whose lives truly hung in the balance. The unluckiest would leave in a hearse.

Some suffered from pre- or posttransplant infections that ate deeper and deeper into their bodies, despite powerful antibiotics and trips to the operating room to flush out the bad stuff. Some suffered from multiple serious afflictions in addition to needing a new organ—hypertension, uncontrolled bleeding, deep depression, untouchable fevers that burned day and night. Some wound up here because fate had frowned on them: they had been born with disease, had contracted it innocently, had been the victim of accident, or had lived unknowingly as one cell after another in some hidden part of them divided uncontrollably until the ugly symptoms of cancer were manifested.

But others had a hand in their plight.

A middle-age alcoholic had received a new liver in 1999 but kept drinking. That liver was now failing, his badly infected pancreas resisted all treatment, and he was diabetic. Clearly he was no candidate for a second transplant—in the era of shortages, no one of sound mind would waste an organ on him. "I think he's going to be a permanent resident of the MGH," Cosimi said. Permanent, in this instance, was a bit of a euphemism.

Someday, Sachs and his colleagues at MGH and other medical centers hoped, even patients like the alcoholic might get another chance.

Despite public education campaigns on the importance of organ donation, donor cards, and stories on TV and in the press, the waiting list for all organs in America grew from 33,014 in 1993 to almost 87,000 in September 2004, according to the United Network for Organ Sharing, the national procurement and matching agency. Two-thirds on the list sought a kidney, with almost 17,500 needing a liver, more than 3,900 in need of a lung, and almost 3,400 in need of a new heart. Thousands would die before getting their chance. Thousands more, like the alcoholic, were not even on a list and were unlikely ever to be.

The shortage also gave rise to a black market in organs: kidneys mostly, since a normal person has two and can live with one, although human rights activists had documented sales of livers, hearts, and other organs from death row prisoners, primarily in China.

In this shadowy zone, brokers connected needy patients from around the world, including the United States, to donors in such countries as South Africa, Brazil, and the Philippines, where poverty is widespread and hospitals have doctors who are trained in transplantation and are willing to skirt the law. Virtually no nation sanctions living donor sales (Iran, where kidney sales are permitted, is one exception) and medical associations around the world condemn the practice as unethical, but the financial incentives to donors, doctors, and brokers are considerable.

And the appeal of transplantation to someone with irreversible kidney failure who cannot survive without dialysis—a time-consuming, restrictive, and often perilous technology—cannot be underrated. Steve Hecht, a longtime dialysis patient who had come to Sachs's attention, said, "Dialysis keeps people alive—and it's probably better than just that, but it's not great, it's really not great. I guess you could call it palliative, but it's a substandard quality of life." Three times a week, for four hours at at time, Hecht, like more than a quarter of a million other kidney patients in the United States alone, is slave to a dialysis machine. Lacking a transplant, most remain shackled for life.

In a 2003 *Los Angeles Times* op-ed piece, Nancy Scheper-Hughes, the anthropologist who heads Organs Watch, which monitors this black market, described a lawyer from Israel. The wait for a cadaver kidney in Israel is several years and the lawyer refused to trust his fate to such an unpredictable system. He told Scheper-Hughes:

"Why should I have to wait years for a kidney from someone who was in a car accident, pinned under the car for many hours, then in miserable condition in the intensive care unit for days and only then, after all that trauma, have that same organ put inside me? That organ is not going to be any good! Or, even worse, I could get the organ of an elderly person, or an alcoholic, or a person who died of a stroke. It's far better to get a kidney from a healthy person who can also benefit from the money I can afford to pay."

The lawyer did not disclose what he paid for his new organ, but estimates by Scheper-Hughes and reporters who have investigated the trade put the cost at up to $150,000, split among broker, doctor, and donor—with the donor often getting the smallest share. Scheper-Hughes says that donors typically receive from $1,000 to $20,000 to give up an organ, with the average only about $2,000. The whole business is a tragedy on many levels, but $2,000 is a fortune to someone living in unbearable poverty, as most donors do.

Alberty Jose da Silva, profiled in a *New York Times* story, was one such donor. The son of a prostitute, thirty-eight-year-old da Silva lived with ten other people in a two-room shack near the airport in Recife, Brazil. "As a child, I can remember seven of us sharing a single egg," da Silva, one of twenty-three children, told the *Times* reporter.

When the man learned that he could earn $6,000 for helping a dying woman—an American, as it turned out—he needed no further persuasion. "Six grand is a lot of money, especially when you don't have any," he said. Two brokers flew the donor and the recipient to South Africa, where the transplant took place. The irony of the operation taking place in the same country where the first human heart had been successfully (and legitimately) transplanted is inescapable.

The woman recipient, who did not give her name, was grateful to da Silva. "I had been on dialysis for fifteen years and on two transplant lists for seven," she said. "Nothing was happening and my health was getting worse and worse." Without a new kidney, her doctors told her, she would die.

A more macabre chapter in live-organ trafficking unfolded in Asia, where the organs of executed prisoners were harvested without the consent of the prisoners or their families.

According to Scheper-Hughes, Asian patients traveled to Taiwan to buy organs from executed prisoners until the World Medical Association condemned the practice in 1994. Scheper-Hughes and other activists claimed, however, that mainland China continued the practice. The anthropologist related a report of a condemned Chinese prisoner who was anesthetized late one night and then had both of his kidneys removed for transplant; the next morning, he was shot in the head.

China's official policy allowed harvesting only when a prisoner or the family gave written consent, or when the body was unclaimed. But Chinese dissidents and human rights groups, including Amnesty International, said the policy was routinely ignored, as tissues and organs, including kidneys, livers, corneas, and skin, were taken at will for buyers—typically wealthy people from Malaysia, Singapore, Taiwan, or Thailand. At a June 2001 hearing before the House Subcommittee on International Operations and Human Rights, deputy assistant secretary of state Michael E. Parmly said that the U.S. government had raised the issue repeatedly to China for a decade.

"The Department of State is also aware," Parmly testified, "of reports that it cannot independently confirm of other, even more egregious practices, such as removing organs from still-living prisoners, and scheduling executions to accommodate the need for particular organs. In addition, there are compelling first-hand reports that doctors, in violation of medical ethics codes, have performed medical procedures to prepare condemned prisoners for execution and organ removal."

A Chinese burn surgeon who had participated in organ harvesting also testified at the hearing. He told of one instance in which a condemned prisoner was shot but not killed. As he lay convulsing on the ground, Dr. Wang Guoqi testified, doctors "extracted his kidneys quickly and precisely." Even then, the man was still breathing, Guoqi testified. "It is with deep regret and remorse for my actions that I stand here today testifying against the practices of organ and tissue sales from death row prisoners," the contrite doctor said.

The practice continues today, according to Scheper-Hughes.

"I just returned from the international transplant meetings in Vienna where I met with Chinese surgeons who sat down with me to ask, 'What is wrong with what we are doing?'" she said in September 2004. The Chinese surgeons believed that if a person was going to be executed anyway, it would be a waste to let precious organs go to the grave when someone living could be helped. It was a tough but not insane notion.

The conference part of rounds ended.

Led by Russell, who with his bow tie and glasses cut an authoritative but kindly figure, the entourage toured the ward, stopping outside each patient's door for a discussion. Some patients slept, while others seemed annoyed at this procession of white-coated dignitaries, many of whom were strangers. Other patients invited Russell and his followers into their room, to answer questions or hear the latest prognosis. Interns and residents were always hovering, but high-powered figures like these did not drop by every day.

Rounds ended and Sachs telephoned Jim Winter, who managed the operating rooms and animal quarters, for an update on Goldie. Winter assured him that everything was going according to plan. Then Sachs joined Russell and Cosimi for breakfast in Mass General's main cafeteria.

Some mornings, Sachs had eggs and bacon. Sachs was raised a Jew and he observed kosher tradition at home. But when he was away, he had no problem eating pork. Today he had a bagel.

Kaz's hands were deep inside Goldie when, shortly before 11:00 A.M., Sachs materialized at the operating room door.

"I couldn't not come!" he said, peeking in.

No one acknowledged him, if anyone even noticed that he was there. Sachs fell silent, and the only sounds again were the suctioning, the mechanized breathing, and the occasional soft exchange in Japanese between Kaz and his assistant. Ordinarily the technicians would have the radio on or would be playing a CD—1970s pop and U2 were favorites—but this was no ordinary day. The atmosphere was solemn. Everyone knew history was being made.

Sachs was still watching when the door opened and surgeon Dr. Frank J.M.F. Dor, a Dutch research fellow, brought baboon 214 into the room.

The animal had been shaved and and sedated, but its

limbs were twitching. Dor laid the baboon on the second table in the operating room, and he and the technicians secured its arms and legs and hooked up the monitors. B214's blood pressure cuff was green and it fit well on the skinny primate arm.

Sheared of most of its fur, the baboon showed an abundance of sinews and muscles but hardly any fat. The animal had black fingernails and dark hands and feet, in contrast to its flesh-colored abdomen. Preparation for surgery had not improved its looks.

A technician placed a mask over B214's face and the animal fell completely asleep. The bellows hit their rhythm, B214 was washed and draped, and Dor began a lateral incision along the baboon's lower belly, where Goldie's heart would soon reside, a few inches below the baboon's own.

Kaz spent much of the morning freeing up the lobes of Goldie's thymus, buried in a bed of fat and tissue in the animal's neck. It was meticulous surgery, but Kaz likely had more experience with pig thymuses than anyone else in animal research—certainly more than anyone at Sachs's lab. As it came into view, the thymus revealed itself to be tiny, crimson-colored, and squishy, like the meat of some rare oyster. The thymus lacked the sanctity of the heart or the gravity of the liver, but it was a vital if still mysterious part of the immune system, and the Sachs group had already demonstrated its potential for helping achieve tolerance.

"Bowl," Kaz said.

He dropped a thymus lobe into a bowl of sterile ice water

and brought it to a table where he and his assistant began to trim. B212, the baboon who would receive it, was already anesthetized and opened in the adjacent operating room. Kaz finished and went next door, bowl in hand.

It was 11:15. They were right on schedule.

With Kaz momentarily out of the way, Cooper and Kawaki stepped up to Goldie. Kaz had done more than harvest a lobe of thymus—he had partially freed up Goldie's kidneys and exposed Goldie's heart. It indeed resembled a human's in color, shape, and contraction: 121 beats a minute, currently.

"Good luck," Sachs said to Cooper.

"Thank-you."

Sachs left, and Cooper and Kawaki prepared to remove Goldie's heart. They were interrupted by Duggan, who wanted Cooper to harvest the pig's ovaries, one of many samples of her anatomy that would be studied.

"They'll be like little berries," Duggan said as he watched Cooper feel around. "There! There! There!" Cooper touched an ovary with an instrument. It was about the size and color of a currant.

"There's the second one," Cooper said.

Cooper snipped the ovaries and dropped them into a bag of ice water that carried a tag identifying the scientist who would study them: "Bob Hawley's ovaries," the tag read.

"I always knew there was something funny about him!" Duggan quipped.

But Robert J. Hawley, Ph.D., was considerably more than

the butt of a lame joke. He was the molecular geneticist at Immerge BioTherapeutics who had knocked out both copies of the sugar-producing gene. He deserved much of the credit for creating the golden pig—but he was not the only one who had helped produce such an animal.

Vectoring

Xeno researchers in Europe, Japan, Australia, the United States, and elsewhere took note on March 15, 2000, when PPL Therapeutics issued a press release claiming to have cloned the first pigs. Whether or not PPL was really first remains controversial—others were known to be very close, including the University of Missouri's Randall S. Prather, who was collaborating with the precursor company to Immerge. But on that day, PPL was perceived as the bold pioneer, at least in the eyes of the public.

According to the press release, five baby pigs had been delivered by Caesarian section on March 5 at the Virginia-Maryland College of Veterinary Medicine: Millie, Christa, Dotcom, Alexis, and Carrel (a nod to 1912 Nobel laureate Alexis Carrel, who developed some of the techniques to suture blood vessels together and laid the foundation for some of the surgical techniques in transplantation). Their birth, PPL predicted, would inaugurate a new era in xenotransplantation. PPL's decision to go to the mainstream press without first publishing in a peer-reviewed publication, the traditional approach, spoke to another of the company's intents: it wanted to reach possible outside investors as quickly as possible. With peer review, months could pass between manuscript submission and publication.

News from the three-page release was distributed by the Associated Press, Reuters, and others, ensuring a considerably larger audience than if PPL's results had quietly appeared in, say, *Mammalian Genome*. Reporters quoted liberally from the release, spinning the story as yet another coup for PPL. This was, after all, the company that owned the magic word in biotech: Dolly. And this was a new era, when scientists courted investors and savvy scientists understood the power of publicity. The line between commerce and medical science was blurring.

"The successful cloning of these pigs is a major step in achieving PPL's xenograft objectives," the company wrote. "It opens the door to making modified pigs whose organs and cells can be successfully transplanted into humans, the only near-term solution to solving the worldwide organ shortage crisis. Clinical trials could start in as little as four years and analysts believe the market could be worth $6 billion for solid organs alone, with as much again possible from cellular therapies, eg. transplantable cells that produce insulin for treatment of diabetes."

PPL explained that due to its unique reproductive biology, a pig had proved more difficult to clone than other livestock; a different method than what was used for sheep or for cows had to be developed. Having now cloned a pig, the company wrote, PPL was already taking the next step:

knocking out the gene, alpha–1,3-galactosyltransferase, that produces the sugar molecule responsible for hyperacute rejection. Further down the line, PPL said, more genetic engineering would be needed to prevent delayed rejection. A means would have to be found to "tolerize" a patient so that long-term rejection did not occur.

On paper, the Holy Grail sounded so simple.

Alan Colman, PPL research director said, "In continuing its proud tradition of achieving world firsts—first to clone an adult mammal, Dolly (with the Roslin Institute), first to achieve gene knock-out in livestock, and now first to clone pigs—PPL has built up the technical expertise and intellectual property to be the first to produce the type of pig which should become the industry standard for xenotransplantation—a pig lacking the alpha 1–3 gal transferase gene."

The release closed with a bold statement: "We are unaware of any other group that has as comprehensive an approach to xenotransplantation as PPL," said Ron James, managing director of PPL. "All the known technical hurdles have been overcome. It is now a case of combining the various strategies into one male and one female pig, and breeding from these. An end to the chronic organ shortage is now in sight."

But two things were lacking—more science, to knock out both copies of the sugar gene, and more money, to keep the program going. "We are looking at various ways to fund our xenograft program," James said, "including discussions with potential marketing partners."

In time, PPL did attract partners that split off PPL's xeno

program and joined it with Starzl's group at Pittsburgh to form a new firm. PPL attracted substantial funding—and even one of Sach's senior researchers—at a time when Sachs himself was struggling. But in March 2000, that was still in the future.

Like others in his field, Sachs found PPL's aggressive marketing a bit distasteful. But the news was nonetheless exciting, whether or not history would prove PPL first in the race to clone a pig. He got on the phone to Elliot Lebowitz, who headed BioTransplant, a Massachusetts biotech firm with which Sachs was then collaborating and the precursor company to Immerge.

"Elliot," Sachs said, "whatever it costs, you've got to get that technology."

But as it happened, they did not need PPL technology. Prather, codirector of the National Swine Research and Resource Center at the University of Missouri's College of Agriculture, Food, and Natural Resources, succeeded in cloning a pig at about the same time as PPL.

The next step for Sachs and his collaborators was the same one that PPL was taking: finding a way to knock out at least one copy of the sugar gene. A cell lacking the gene could then be used to clone a knockout pig.

In fact, by the time of PPL's cloning announcement, Hawley had already knocked out one copy of the gene.

Hawley brought impressive credentials to the job. He had earned a Ph.D. in developmental and molecular genetics at Marquette University with a thesis entitled "Molecular

Genetics of Oogenesis in Drosophila"—fruit flies—then spent four years as a postdoctoral fellow at Boston's Dana-Farber Cancer Institute, where he researched the cytokine regulation of neural gene expression. Hawley joined Bio-Transplant in 1992, two years after the firm was founded, and rose quickly to the position of principal scientist.

All pigs are born with two copies of the sugar gene: one from the mother and one from the father. In attempting to get rid of the first copy, Hawley used gene-targeting technology, in which a piece of DNA is introduced into a cell and directed to a site on the genome—which it then replaces, or knocks out. In the case of the sugar gene, the vector was inactive DNA whose only effect was to remove the alpha 1–3 gal gene. Gene targeting is what Hawley described as a "low frequency" procedure: starting with as many as 10 million cells, only one may be successfully targeted, making the research time-consuming and laborious.

But Hawley persevered.

By 1998, two years before PPL announced that it had cloned a pig, Hawley had single-knockout cells from David Sachs's pigs. They were useless, though, until someone on Sachs's side figured out how to clone a pig.

In September 2001, Prather succeeded in cloning pigs that lacked one copy of the sugar gene. A second litter was born that October. BioTransplant's xeno technology by then had been folded into Immerge, and Hawley, who held the title of associate director of animal genetic engineering with the new firm, opened bottles of champagne with his colleagues. Prather began to inbreed the single-knockouts, with the expectation that the normal laws of genetics would hold and pigs lacking both copies of the gene eventually would be born. An account of the September and October births and the events leading up to them was submitted to *Science* and, after peer review, was published online on January 3, 2002, with Hawley, Prather, and others named as authors.

Theirs was a solid, respectable paper, written in the arcana of molecular genetics and including photographs, charts, a diagram, and an extensive bibliography. It ended with an optimistic, if clumsily worded, assessment of the future of xeno: "We hope that a-1,3-galactosyltransferase-null pigs will not only eliminate hyperacute rejection but also ameliorate later rejection processes, and (in conjunction with clinically relevant immunosuppressive therapy) will permit long-term survival of transplanted porcine organs."

But Hawley and Prather had been scooped. The day before their *Science* article was published, PPL issued another press release. There would be no controversy this time about who had been first—it was Hawley and Prather—but PPL nonetheless managed to steal their thunder, again by forgoing the peer review process. Once again, their marketing was as good as their science. And the pigs' birthdate, December 25, provided them the opportunity to personalize the animals by giving them cute Christmas names. (Hawley and Prather referred to their animals only as "piglets," as most scientists would have done.)

"World's First Announcement of Cloned 'Knock-Out' Pigs," the headline to PPL's release read. "Christmas-born pigs are a major step towards successful production of animal organs and cells for human transplantation use."

Using the same confident tone as its March 2000 statements, PPL went on to proclaim the birth of Noel, Angel, Star, Joy, and Mary, another bold step toward xeno's bright—and, for those investors who might be paying attention, profitable—future. The company was aggressively seeking to sell its xeno business so that it could focus on developing its recombinant human alpha1-antitrypsin, or recAAT, a substance produced from the milk of genetically altered sheep; used to treat emphysema, recAAT was already in clinical trials. The shift in focus suggested to some that PPL was not as confident of the prospects for commercial xenotransplantation as it maintained.

"Today's announcement is a natural breakpoint for PPL to spin out the valuable technology it has developed thus far," managing director James said. "In light of this news, finding a third party at this particular time to take forward this very exciting area of science, which addresses major markets, will ensure that PPL's shareholders gain maximum value, whilst protecting the Company's limited cash resources needed to bring its lead product, recAAT, to market as quickly as possible."

Beaten to the punch, Immerge nonetheless got some publicity by issuing a release correctly taking credit for being first with single-knockouts. But the emphasis was on the advantages of Sachs's unique line of pigs (PPL used a different type), from which the knockouts had been cloned. With Novartis's funding seemingly secure, Immerge executives felt no compulsion to follow PPL's marketing tack.

"We have been actively developing a line of miniature swine that offers many advantages as a potential donor for xenotransplantation, including their organ size, which is appropriate for human recipients," said Julia L. Greenstein, chief executive officer and president of Immerge.

"Based on the success reported today, we can now proceed with pig strains chosen solely for their advantages in xenotransplantation rather than their large-scale availability," Hawley added, in a subtle knock on PPL, which had not produced its clones from Sachs's herd. "We believe this line of miniature swine offers the greatest potential for clinical use in humans."

As Prather continued to breed his single-knockouts, in pursuit of pigs lacking both copies of the sugar gene, Hawley was back in his lab, trying to knock out the second copy; if there was any way around it, Immerge did not want to wait for the breeding program to bear fruit. That might take two years or more, with no guarantees of success. Immerge—and Sachs—needed double-knockouts as soon as possible. PPL was their chief rival, but there were other scientists trying different strategies at other centers, notably the University of Minnesota and the Mayo Clinic College of Medicine, which received support from Baxter, the $9 billion health care company.

Hawley tried using the technique he had used to remove the first copy of the gene, but it didn't work on the second copy, for reasons he never unearthed. "Still don't know why," he said several years later. "It was very frustrating." Other researchers tried similar techniques and were also unsuccessful. Much remained to be learned in genetic engineering.

Hawley then turned to Prather's single-knockout pigs. He harvested cells from single-knockout fetuses and piglets and grew them out, by the millions, in the lab. Hawley was able to isolate a small number of cells that, through spontaneous mutation, lacked the second copy of the gene. Hawley grew out a line of them and sent some to Prather and also to DeForest, Wisconsin–based Infigen, another biotech firm with pig-cloning capabilities with which Immerge collaborated.

But PPL won this race uncontestably. On July 25, 2002, months before Prather and Infigen succeeded, four pigs lacking both copies of the sugar gene were born in Virginia.

In typical fashion, PPL (which now billed itself as "one of the leading biopharmaceutical companies in the application of transgenic technologies") published the results in a press release on August 22. It noted that the company was now collaborating with the University of Pittsburgh's Thomas E. Starzl Transplant Institute, where researchers would conduct experiments to determine if the double-knockout pigs' organs and tissues lived up to their promise. PPL also said it expected to sell its xeno unit by the end of the year.

Prather finally broke through on November 18, when Goldie was born in Missouri. Immerge's other cloning partner, Infigen, soon followed with a litter of three piglets lacking both copies of the sugar gene.

Sachs monitored the pregnancy that resulted in Goldie with caution. "I didn't get my hopes up," he said. "This is a field where most of the pregnancies terminate in spontaneous abortion."

Even with Goldie's birth and the subsequent determination that the animal was healthy and suitable for experimentation, Sachs had restrained his emotions. Keeping the piglet alive and getting it safely to Boston were fraught with hazards. And there was still the larger point: would a double-knockout pig really make a difference in xeno?

"I've learned over the years," Sachs said, "that if you let yourself get too excited, the disappointment is too great."

Hawley, Greenstein, and the rest of Immerge took a similar position when Goldie was born. It wasn't until Hawley flew to Missouri a short while after the animal's birth and returned with photographs of the piglet that the staff celebrated with cake and champagne.

Immerge was similarly cautious in announcing Goldie to the outside world. The scientists waited until January 13, 2003, when Prather spoke in Auckland, New Zealand, at the annual meeting of the International Embryo Transfer Society. The city and the society lacked the pizzazz of a PPL announcement, but this time, Immerge issued a press release that signaled excitement. Greenstein went a step further with word that Sachs's pigs—and pigs cloned from them—apparently harbored a harmless variety of a pig virus that troubled some scientists and investors. Porcine endogenous retrovirus (PERV), as it was called, resided on pig genes—with no effect to the pig. But so far, no one had been able to rule out the possibility that it could act differently in a human host, transforming itself into some terrible new disease that could spread into the human population, much as AIDS had done.

Greenstein knew she had to sell safety.

"The strain of swine we are working with seems to be incapable of transmitting porcine endogenous retrovirus to human cells in culture, as we reported in March 2002 in the *Journal of Virology*," she said. "Although the risk of any harm posed by PERV to xenotransplant recipients may be purely theoretical, use of this line of miniature swine would help minimize this particular risk of this new technology."

Immerge did not disclose when experiments on Goldie would begin or what they would entail. Still, major media outlets ran the double-knockout story, and the accounts brimmed with optimism. Contacted by the *Boston Globe,* Cooper said, "The immune response to this sugar has been the major problem facing us for the last ten years. Despite our best efforts, it's not been possible to get over it completely. Now it is." A magazine at Prather's school wrote, "She may not be as famous as Arnold, Babe or Wilbur, but . . . Goldie has accomplished something her well-known peers never thought of doing: offering hope to more than 80,000 Americans currently awaiting organ transplants."

Even the Urbandale, Iowa–based National Pork Producers Council, which bills itself as the global voice for the U.S. pork industry, hailed the double-knockouts. (The council's Web site offers such illuminating observations as this: "Pigs are often thought to be dirty, but actually keep themselves cleaner than most pets. They are seen laying in mud because they do not have sweat glands and constantly need water or mud to cool off.") The headline on a pork council release announced, "Pigs Could One Day Shorten the Wait for a Transplant."

Staying Cool

At noon, Cooper instructed technician Crystal Dugan to infuse Goldie with heparin, which would prevent blood from coagulating and gumming up the heart's four chambers once they had been stilled. Ten minutes later, Cooper finished sewing a tube into an artery running directly into Goldie's heart.

He asked Dugan to administer a dose of cardioplegia, a mix of chemicals that temporarily stops beating.

"It's running," Dugan said.

"Running well?"

"Running well."

"Cold saline," Cooper said. He poured two bowls of it directly onto the heart, and it gave up its beat in seconds. Dugan turned the ventilator off. An alarm beeped. "Equipment alert. No pulse detected," the monitor read.

Kawaki cut the major vessels and lifted the organ out.

B214 was waiting.

Goldie was gone.

Although her case did not come up at rounds on the Wednesday Goldie's organs were taken, Cheryl Snow, a forty-three-year-old mother of four who lived in Revere, Massachusetts, was awaiting a transplant in a room at Mass General Hospital's main campus.

She needed a new heart, and someone would have to die for her to get it.

Snow was the sole occupant of her room. Visitors were asked to call ahead and had to wash their hands with germ-killing soap before entering. Snow herself could not leave, save for occasional walks up and down her ward and trips to the catheterization lab, where a line was threaded into her dying heart. Periodically it had to be replaced, and that hurt.

"I want to get some fresh air but I can't go outside—

they're afraid I'll get an infection," Snow said. Infected, she
would be taken off the waiting list until her illness cleared,
assuming it did.

Snow had been healthy until 1998, when she began to
have trouble breathing.

"I'd be walking up the hill to my son's school and I
couldn't make it, I was so short of breath," she said. "At that
time I smoked, so I quit."

Her condition worsened.

A doctor diagnosed asthma and prescribed medication,
but that did nothing: just crossing a room left her winded.
She returned to the doctor, who ordered an EKG and then
pronounced her fine.

Snow started crying in the office. "I know there's something wrong with me!" she said. "I can't breathe!"

Snow went to Mass General, and in March 1999 doctors
diagnosed dilated cardiomyopathy, a disease that causes the
heart to swell and pump ever more weakly; then the lungs
fill with fluids, making breathing a chore. The cause is often
unknown, but viruses can be at fault—and there may be a
genetic predisposition. That seemed likely in Snow's case,
since her father, maternal grandmother, and paternal uncle
all died suddenly, at age fifty-seven, forty-two, and fifty-one
respectively. According to Cheryl, they all were victims of
heart attacks.

Although cardoiomyopathy is incurable, some patients
can lead normal lives with medications. Snow was one of
them—for a spell. After ten days in the hospital, she went

home, and while she tired more easily, she was healthy enough to return to her job at a Home Depot store. But her health did not last. In August 2002, while she was at work, she went into cardiac arrest. But for the chance presence of a shopper who knew cardiopulmonary resuscitation, she would have died.

Back at Mass General, the doctors implanted a pacemaker and a defibrillator. She returned in October for a new pacemaker and her cardiologist, Dr. Thomas G. DiSalvo, told her that she needed a new heart. She was placed on the waiting list immediately.

"I went back home and I said to myself, All I want to do is make it through Christmas," Snow said. "Because I knew I wasn't going to last long. I just wanted to spend Christmas at home with my family."

Snow's condition deteriorated over the holidays, and by January 19, DiSalvo felt it safest to admit her. She would remain hospitalized until she reached the top of the list.

A month later, she hadn't reached it. There was no guarantee she ever would.

She passed the days reading, listening to CDs, watching TV, crocheting, and entertaining visitors, including her husband, who came by nearly every day.

"My husband is like, 'I don't know how you do it,'" she said. "And Dr. DiSalvo told me that some people commit suicide—they can't deal with this. I have a pretty good mind, I guess. That's what everybody says: how calm I am. I've always been like that: very relaxed, laid-back."

Psychological testing confirmed that Cheryl could handle the pressure of being a transplant candidate and, God willing, a recipient. For a kidney or liver, it is sometimes possible to find a living donor (a segment of liver is used in living-donor transplants), but a transplanted heart always results from a tragedy. "You know that your only hope is someone else getting killed in an auto accident," Sachs said. "It's not pleasant. It's obvious you didn't do anything to cause it, but nevertheless you're the beneficiary."

With tolerance still only a vision for cardiac patients, the staff also needed assurance that Snow could keep track of the more than dozen medications she would have to take daily for the rest of her life. They needed to know that she would return regularly to the hospital for monitoring. They had to be sure that she did not abuse drugs or alcohol.

They wanted someone for whom the act of living was precious.

"There are so many people waiting for organs that they don't want to give it to a person that's gonna just go out and destroy themselves," Snow said. "It's their way of deciding who's going to get on the list and who's not going to get on the list. Some people don't make it. And some people are on the list and they get taken off."

Snow's equanimity masked an impatience that, as the weeks unfolded, would intensify. If xenotransplantation had been perfected, she said, she would not hesitate to accept a heart from an animal.

"Would I use a pig organ? If I knew it was going to work,

definitely. I'm sure everyone else would too. There should be an abundance of organs. They shouldn't even be taking people off the list or having to do anything to be put on a list—just everybody that needs one should be able to get one."

Cooper and Kawaki worked like fussy tailors on Goldie's heart as it rested in ice water. They wanted the smoothest possible fit inside B214. They did not want technical mistakes to ruin their experiment.

"Trim that edge a bit," Cooper said. "Get rid of that artery; we don't need it. Take a little bit more here."

Satisfied, Cooper and Kawaki carried the chilled organ to the sleeping baboon. Kaz, meantime, had completed transplanting the thymus lobe into B212 and he was back with Goldie, set to transplant the kidney and other thymus lobe that he would put into B113, which had been anesthetized.

Other surgeons and assistants were taking more samples from the still-warm pig: slivers and slices of pancreas, ear, liver, lymph nodes, tonsils, skin, stomach, large and small intestine, salivary glands. "This is not the fun part," said Dutch fellow Dor, assigned to take bone marrow, which necessitated sawing and cracking open bones. He said it reminded him of the butcher. Dor had a fine sense of humor—he wore a pig costume to a Halloween party—but this part of his job depressed him.

It was a few minutes past one o'clock. They had fallen

almost an hour behind schedule—an acceptable slip, given all that was happening.

Cooper removed the moistened bandage protecting B214's insides and felt around inside the abdominal cavity. Goldie's heart would not replace the baboon's; the purpose of this experiment was to study the immunology of a double-knock-out organ, not determine if a pig heart could sustain a baboon's life. And so it would be a redundant heart—a so-called heterotopic heart—a secondary organ of no value to B214, only to medical science, nestled inside the baboon in an area near the intestines where major vessels would provide circulation. Even if Goldie's heart failed or was rejected, B214 would still have its own and might live to see another day.

Cooper and Kawaki needed less than an hour to connect the heart and unclamp the vessels, allowing the baboon's blood into it. So far, there was no sign of hyperacute rejection. But it was too soon to be ruled out.

"Okay, let's have a look at this, Kenji," Cooper said. "Perfect. Okay. Good. No bleeding from the suture line. Kenji will get paid this week!"

The phone rang and Winter answered. It was Sachs, calling from the hospital's main campus. He could have been an expectant 1950s father, eager for word of how his wife's labor was unfolding in a room he was forbidden to enter.

"No problems whatsoever," Winter said. "It's run quite smoothly actually." In fact, he said, one baboon was already back in its cage.

"I'm really pleased!" Sachs said.

"Let's have the paddles," Cooper said.

Winter switched on the defibrillator and handed Cooper the electrodes. Cooper held them on opposite sides of Goldie's heart.

"Okay," he said.

Winter hit the button but the heart did not budge.

"Try again, Jim."

"Twenty?"

"Twenty."

Winter turned the power up.

The second shock made the atria quiver, but the other two chambers didn't budge.

"Okay, we'll try again, Jim," Cooper said.

"Whenever you're ready."

The third zap failed, too.

"Try one more."

The fourth shock lifted the unconscious B214 off the table.

Goldie's heart began beating properly.

Baboon blood coursed through it, and Cooper watched with satisfaction, remembering what hyperacute rejection looked like: within a few minutes, as you watched in growing disappointment, the transplanted organ turned a dusky hue and died. The immune system was mighty indeed.

"I remember thinking, Are we ever going to get over this? This is such a violent thing," Cooper said.

But this double-knockout heart was a beautiful pink.

*

In his background and in his demeanor—and even in his physical appearance—Dr. David Kempton Cartwright Cooper stood in contrast to Kaz.

The British-born researcher was tall—at six foot five, he towered over his Japanese colleague and everyone else on Sachs's staff. He counted the writers Michael Crichton and Robin Cook and stem cell researcher Lanza among his friends. He had a literary touch, not only in the style of the hundreds of scientific articles he'd produced, but also in the novel, short stories, and plays he had written. He had rowed at university, a fact that he included as the third paragraph of his official abbreviated curriculum vitae: "As an undergraduate, he captained the Guy's Hospital Boat Club, and as a postgraduate, he rowed in the University of London 2nd VIII (when the 1st VIII was the British Olympic crew), and represented Wales in the Home Countries International Regatta, winning the coxed IV event."

The son of a businessman and what he described as "an old-fashioned homemaker," Cooper was born in 1939 in London. He was to be given only one middle name—Kempton, after his father, William Kempton Cooper—but Cartwright, another old family name, was tacked on at the last minute. "My grandmother was present at the christening and was upset that Cartwright was not to be included," Cooper said. "My father added it to please her." Nonetheless, Cooper used only K.C. in his correspondence.

No one in his family practiced medicine, but Cooper

became interested when he was a teenager and toured a hospital near London. He graduated from Guy's Hospital Medical School of the University of London in 1963, spent a year teaching anatomy at Harvard Medical School, and then returned to England to earn a doctorate from the National Heart Hospital and Institute of Cardiology in London. He was a young man with wanderlust.

Influenced by the noted British heart surgeons Lord Russell Brock and Donald Ross, who had experimented in xenotransplantation, Cooper decided on cardiac surgery. Barnard's work also intrigued Cooper, and in 1980 he joined the staff of Groote Schuur Hospital in Cape Town, South Africa, where Barnard had performed the first successful heart transplant. Barnard was virtually retired, and Cooper took over the heart transplant program.

Cooper was in South Africa in 1977 when he became interested in xenotransplantation, which Barnard had attempted twice. Barnard's first patient, recipient of a chimpanzee heart, died after four days, and the second, recipient of a baboon heart, died after just six hours. Cooper began transplanting pig organs into baboons in 1984, years before Sachs started such experiments. His motivation, like Sachs's, was the increasing shortage of organs. Xenotransplantation seemed a logical solution.

"I was first attracted," he said, "by the realization—when running the heart transplant program in Cape Town—that we would never have enough human donor organs."

Motivated in part by the chance to make more money,

Cooper left South Africa in 1987 for the Oklahoma Transplantation Institute in Oklahoma City, where he continued his xeno research and the practice of human heart transplantation. While in Oklahoma, Cooper, working with others, identified the sugar molecule that caused hyperacute rejection of pig organs. This was the groundwork for the genetic engineering that created Goldie and other double-knockout pigs.

Cooper stayed in Oklahoma for nine years and then his wanderlust returned. He started looking around for something new.

"I always think a change—a new challenge—is a good thing," Cooper said. "It wakes you up and gives you a new sort of spurt of energy. You either need a change of geography or you need a change of job description. In my life, I've done this every seven or eight or nine years."

Cooper was offered a professorship in Sweden—but then Sachs heard of the surgeon's desire to move. Sachs asked if he wanted to devote all of his time to research, and Cooper said that he did. "After seventeen years in clinical heart transplantation, during which time research had become increasingly important to me," Cooper said, "I decided I would give up clinical work as this commitment made it difficult to concentrate on research." Cooper moved to Boston in 1996, joining the staff of Sachs's Transplantation Biology Research Center and becoming an assistant professor at Harvard Medical School.

The early work in pig-to-baboon heart transplantation was discouraging. "When I started," he said, "we were lucky to get a week survival—a week or two. And we were giving a lot of treatment, and it was a huge procedure. We were doing plasma pharesis and removal of antibodies. We were giving irradiation, we were giving big doses of immunosuppressives—all this sort of stuff." In the mid-1990s, before Goldie, Cooper couldn't predict when—or if—xenotransplantation would become a clinical reality.

At 2:20 P.M., as Cooper adjusted B214's medications, Goldie was wheeled into the other operating room, where Dor and an assistant would continue to harvest bone marrow—this batch to be transplanted by catheter into another baboon later in the day. Soon there would be precious little of the golden pig left.

Cooper was concerned by the pace of Goldie's heart inside its new host, a baboon. It was beating about fifty times a minute, irregularly.

"I can't understand why this heart isn't picking up," Cooper said. "His own heart isn't going up, either."

Cooper ordered the heat up in the room and increased the dopamine, a stimulant. B214's heartbeat increased to 120, Goldie's to 80. Cooper rechecked the suture lines and removed the sponges, providing an unobstructed view of the heart. He asked an assistant to take a picture, and then he and Kawaki began to close the wound. When they were done, shortly after 4:00 P.M., the steady beat of Goldie's heart was visible beneath the skin over B214's belly.

"You can wake him up," Cooper said.

The technician shut off the anesthesia. B214 quivered momentarily but did not come to. The animal's native heart seemed to be in trouble, with its beat fluctuating wildly: a pulse of 100, then 88 a few seconds later, then 110 a few seconds after that. Cooper wiggled the animal's left arm while Winter stroked its face and ear, to no response.

"He's got rather large, dilated, fixed pupils, which worries me," Cooper said.

The pulse dropped from 77 to 51 in less than a minute.

But suddenly B214's right leg began to move. Its pulse and blood pressure stabilized and then began climbing, and then Meaghan Sheils arrived at the animal's side.

Sheils held the baboon's hand and stroked its face.

"Wanna wake up?" she said, her voice babyish but soothing.

The animal remained still. Sheils scratched its ear and head and tickled his nose and tongue.

"Put your silly tongue in!"

The animal stirred.

It was 5:34 P.M. Anesthetized for over six hours, B214 was coming to.

"Okay, you can disconnect him," Cooper said.

They sat the animal up on the table. It looked like a child emerging bewildered from a dream.

"Hi, handsome!" Sheils said. "Say, 'That's what I always look like.' You're cute though, huh? Very cute!"

"You can take him back now," Cooper said.

Kawaki carried the animal out of the operating room and across the hall to the baboon room, where a fresh lambswool had been placed in his cage. B214 went slowly to his perch and sat quietly. He took in the activity of the room and watched intently as pain medication was administered. B214 was becoming himself again.

"I think he's okay," Cooper said.

The surgeon offered his expectations for the experiment.

"I would hope to get at least three months," he said. "If we go much less than that we're not doing much better than the old ones. But remember, predictions are risky—especially about the future!"

A short while later, Winter telephoned Sachs.

"Everything went very well," he said. "All of the baboons are back. All are sitting up. All are perched. So they look quite well."

"I'm elated!" Sachs said. "That's the call I've been waiting for!"

The technicians cleaned the operating rooms and what remained of Goldie was placed in a plastic bag that would be stored in a freezer until a company that incinerated medical waste took it away.

"There's really not much left," a German research fellow commented. "Did you take the brain?"

"No," said Duggan. "Nobody wanted the brain."

CHAPTER 3

Waiting Rooms

Miss O'Shea

With his persistent fever and headache, David Sachs was thought to have nothing more than a bad cold that summer of 1946—until he tried to get out of bed and couldn't stand because his left leg had gone limp. Doctors diagnosed paralytic polio, an incurable disease that evoked images of wooden wheelchairs and iron lungs, coffin-like breathing machines that encased a patient's body, save for the head.

David was four and a half years old.

The Sachses had just moved from Washington Heights, New York, to Yonkers, which in the immediate postwar period still had its share of trees and open spaces. David's father, Elliot, ran a jewelry store in Brooklyn, while his mother, Elsie, was home with the three children: David, his older brother, and his younger sister. Descendants of Jewish immigrants who left Eastern Europe in the late nineteenth century to escape pogroms, the Sachses claimed their piece of the American dream.

Now their son had fallen victim to a terrible epidemic. Elsie cried, telling relatives and friends. David was so ill and, furthermore, in 1946 no one fully understood how polio was transmitted. Victims often became outcasts, like early AIDS sufferers.

Sachs was admitted to Manhattan's Hospital for the Ruptured and Crippled, where treatment for patients whose lungs were unaffected included physical therapy and spinal taps to analyze spinal fluid. The little boy screamed and struggled when they came at him with their needles, but the grown-ups always prevailed.

"I remember the lumbar punctures, where one person would hold my arms and legs and the other would be putting the needle in my back," Sachs said. "I just remember them so vividly, even to this day."

Sent home in a wheelchair and with the possibility he would never walk again, the little boy with the lifeless leg and odd-size foot spent weeks near a bay window in the dining room, where his parents had moved his bed. This brought David close to his family and eliminated the need to get up and down the stairs—and it gave him a view of the children at play outside. None dared visit the convalescent, and they shunned his siblings too.

"In those days," Sachs said, "everyone thought, 'Stay away from those children.'" In reality, David was not contagious, and his brother and sister had probably acquired immunity. But it would have taken unusual prescience to understand that, given the many unknowns about polio in 1946.

David was rehospitalized for a period, and when he returned home, this time for good, a physical therapist named Miss O'Shea came by every day. She massaged his leg and showed him how to maximize use of the muscles that still responded. She helped him from his wheelchair onto his feet—moving him across the floor as she repeated her slogan: "Walk away with Miss O'Shea!"

Within a year, Sachs was walking again, but he was embarrassed by the evidence of his disease. One day decades later he sat in his office and lifted the cuffs of his trousers. "See that?" he said. "My feet are two different sizes. I buy two different sets of shoes and have to split them. You wouldn't know it unless I told you, but they're different. When I was a kid I was ashamed of that. I didn't want anyone to know it so I didn't tell anybody. Just didn't want to be different, like all kids don't."

Born at 2:38 P.M. on January 10, 1942, in a Manhattan hospital, David Howard Sachs as a child was fascinated by both nature and things mechanical. He liked photography—and he developed his film and printed his prints in a darkroom that he built. He experimented with his chemistry set. He broke down and rebuilt car engines and was good enough at repairing home appliances that his mother nicknamed him

"Mr. Fix-it." He took care of the family pet, a cocker spaniel named Brandy. And he gardened.

"I was the only one in my family who did, no one knows why," he said. "I turned over a portion of the backyard and had the best flowers and vegetables. Still today when I see a seedling come up it just gives me a feeling of excitement. Every time I've moved to a new house the first thing I've done is gone out back and figured out where I was going to have my vegetable garden, and turned over the soil and planted my seeds. I just love gardening—every spring to see the renewal."

Despite his polio, Sachs did not miss a grade in school.

He was an exceptional student, and his report cards from Southeast Yonkers Junior-Senior High School show A's in nearly every subject. Schoolwork came so easily to him that Elsie and Elliot worried that he lacked ambition because they never saw him studying—he did his homework at school. "My parents—my mother in particular—thought I was never going to amount to anything," Sachs said. "I was always running out to play." They compared him to his brother, Coleman, who was a model student. But Coleman, unlike his younger brother, did not have a photographic memory. David could read a book and still see the words on the pages years later.

Sachs was leaning toward a future in medicine when the Russians launched Sputnik, the world's first satellite, in the autumn of 1957, when Sachs was in eleventh grade. The dawning of the Space Age convinced him that he wanted a career in physics, and he enrolled at Harvard College in

1959, declaring that and chemistry as his majors. He also joined the Harvard swim team, continuing the sport he'd started at Yonkers High School, when he was a kid determined to demonstrate that polio was behind him.

Sachs was a sophomore when he took his first course in organic chemistry, taught by Louis F. Fieser, a scientist who counted among his achievements the development of drugs to treat malaria and cancer, and who belonged to the U.S. surgeon general's committee that in 1964 linked cigarette smoking with disease. But during the 1960s Fieser was vilified as the inventor of napalm, which he and colleagues at Harvard developed in 1943 for use by the Allies in World War II. Fieser later said that because his country had asked him to, it had been his patriotic duty to develop the incendiary weapon. "I have no right to judge the morality of napalm just because I invented it," he told a magazine reporter in the late 1960s. This was not a philosophy that Sachs, who held liberal views, would have shared.

But Fieser's politics did not draw Sachs to the man. The Harvard professor was a brilliant scientist and a mesmerizing lecturer—an animated, imposing character who excited the young student. "I used to run to get [to his class] to get a front seat because it was like going to a good show," Sachs said. Sachs switched his major to chemistry only and graduated summa cum laude from Harvard in 1963. Encouraged by Fieser and supported by a Fulbright fellowship, Sachs spent a year at the University of Paris earning the equivalent of a master's degree in organic chemistry.

But the year convinced Sachs that his passion for chemistry had more to do with Fieser's charisma than with the science. When he entered Harvard Medical School in 1964, his interest shifted to surgery. He was a sophomore when a professor lectured on the emerging field of transplantation. The professor mentioned the British immunologist Sir Peter Medawar, who shared the 1960 Nobel Prize in physiology and medicine for his pioneering work in demonstrating natural tolerance in a rare kind of cow. Medawar's work would become the basis for later research, including Sachs's, into tolerance for people.

"It just blew my mind," Sachs said. "The possibility of being able to give people a new life when they're dying because an organ is failing—and doing it without having to give them drugs that could kill them—it just seemed to me that this was very important."

While Sachs was still in medical school, his roommate told him about a nurse he had seen at the hospital. "You ought to meet her," said the roommate, who was already engaged.

"I'd love to," said Sachs.

Sachs was in the lab, working on one of Paul Russell's projects, when the roommate walked in with Kristina Olsson. Kristina was uncommonly pretty, and she had stylish taste. Perhaps it was in the genes: her aunt was Lisa Fonssagrives, a Swedish-born model of exceptional beauty who graced the covers of fashion magazines during the 1940s, 1950s, and 1960s.

"There I was operating on some mice and she came in!" Sachs said.

Sachs asked if she'd like to go for a cocktail, but Kristina didn't drink. "Okay, let's get a hamburger," Sachs said. They rode to the restaurant on his motor scooter.

The two became a couple, and in 1968, while David was serving his internship and Kristina was pursuing a master's degree in nursing education, David proposed. They married in 1969 and planned to start a family.

The next year, Kristina became pregnant. The Sachses told family and friends the good news—and then Kristina's stepfather died while Sachs was at a transplantation conference in the Netherlands. David booked a flight home while Kristina boarded a train for New York, where her mother and stepfather had their home. Kristina was on the train when she began to miscarry.

"I figured there was a lot of stress going on," Sachs said. "I said, 'We'll get over it and we'll have another baby, another pregnancy'—which we did very soon thereafter. She got pregnant again—and had another miscarriage. After that, we started to get very worried."

When Kristina became pregnant for the third time, her obstetrician prescribed hormone therapy and ordered her to bed. But once more, she did not carry a baby to term. "We were devastated," Sachs said. "It was the worst time of our lives." With Kristina apparently incapable of bearing a child, the couple decided to adopt. An agency that brought children to the states from Korea identified a woman who was giving up her baby—a boy whose father was an American soldier. David and Kristina named the child Erik, and they announced his imminent arrival to relatives and friends. They outfitted a baby's room and put photos of Erik on the walls. And then the mother changed her mind and kept her child after all.

Kristina, meanwhile, had become pregnant again. This time, Sachs was optimistic. "I just had this feeling that something was different," he said, "and I said that to Kristina. I remember hugging her while she was crying and saying, 'Kristina, I think this pregnancy is going to be okay.'"

Seven months later, in May 1973, Michelle Bess Sachs was born. Jessica Allyson, Karin Danielle, and Teviah Erik followed. "In four and a half years, we had four normal, wonderful babies," Sachs said. "There's not a day that I don't think of how blessed we are."

Sachs graduated magna cum laude from Harvard Medical School in 1968 and served his internship at Mass General. The year he married, he began a surgical residency and research fellowship at Mass General. He read all he could about transplantation and learned about xenotransplantation, including the work of Tulane University surgeon Dr.

Keith Reemtsma, who in 1963 and 1964 transplanted chimpanzee kidneys into thirteen people. Of all animals, chimpanzees are closest to people; they share more than 98 percent of their DNA, which provides a xenotransplanter an immunological advantage.

Reemtsma's first case was forty-three-year-old dock worker Jefferson Davis, who was dying of kidney failure. Citing the difficulty in finding human kidneys and the stigma still attached to removing organs from brain-dead people, Reemtsma bought an eighty-pound chimpanzee from a circus whose owners were tired of its bad temper and transplanted both of the animal's kidneys into Jefferson. He was alive and well six weeks later, and Reemtsma agreed to bring him to a press conference.

"I feel better now than I have in five years," Jefferson said. "I was worried, of course, about the animal business. I knew it would be a monkey. It didn't bother me. All I wanted to do was survive. I feel wonderful."

Three weeks later, Jefferson died—of pneumonia. Reemtsma tried again several times, with results ranging from nine days' survival to a schoolteacher who lived nine months with chimpanzee kidneys—dying not of rejection but an acute electrolyte imbalance.

The work of Reemtsma and other researchers was another epiphany for Sachs. "I already had the idea that we were going to need another source of organs besides humans," he said, "that the need for organs was going to exceed the availability of donors. I saw no reason why the organ of an animal—other than the brain—would not be as good an organ as the organ of a human being."

But few scientists were looking at the animal Sachs had in mind.

A Major Complex

Sachs did two years of a surgical residency at Mass General and then took a research job at the National Institutes of Health. He planned to someday complete his residency—but he never did. He soon was chief of the Transplantation Biology Section, Immunology Branch, at the National Cancer Institute.

Sachs sought to further knowledge of the immune system and the mechanisms that govern rejection of a conventionally transplanted organ—an organ from one person put into another, allotransplantation. Knowledge would lead to improved protocols and hopefully clinical tolerance.

Using mice, rats, and other animals, scientists identified the cluster of genes in mammals that played a major role in rejection in same-species transplants: the major histocompatibility complex, or MHC, as it was called. The MHC genes controlled the types of protein markers—antigens— that appear on the surface of each of a person's trillions of cells. Except for identical twins, each individual has a unique set of antigens and can make antibodies that react with foreign antigens when tissue is transplanted from another person (or animal). In such a case, the antigens activate the immune system to destroy the invader. Evolution intended the mechanism as a defense against germs, not as a challenge to life-saving transplantation.

In 1971 Sachs was committed to improving human allo-transplantation, still in its infancy. As he looked for the right animal model for his research, he was already imagining the day when he could begin experiments on animals whose organs might be transplanted into people. Mice and rats, of course, were too small. Which animal would best suit both fields? he wondered.

Sachs did not broadcast his interest in xeno. "Everyone would have thought I was a little crazy!" he said. "This was sort of my own interest." In 1971 xeno was fringe science: The money and excitement were in allotransplantation, which was a clinical reality, albeit one with many risks and unanswered questions.

Although primates possessed genetic similarity to people, other characteristics, Sachs believed, made them less than ideal for xenotransplantation. Chimpanzees were an endangered species—and too close to their human cousins in appearance and even some of their behaviors to ever be accepted by the public as a source of organs. Taking a heart, liver, or lung meant killing the animal, which would have struck many as morally offensive.

Baboons did not have cute appeal. But baboons and monkeys could harbor many diseases, including several that were fatal and were easily transmitted to people. And those were the known diseases. Domesticated animals had lived side by side with people for centuries, but no one knew what germs creatures from the jungle might carry. A decade before AIDS and ebola, Sachs didn't warm to the possibility.

Sachs considered baboons less than ideal for another reason: they took years to reach their full size, and they rarely grew bigger than about sixty pounds. Consequently their organs, the heart especially, would be unable to support a large adult human. Female baboons deliver babies in ones or twos, which would make breeding them to satisfy human demand an expensive, time-consuming proposition.

One day Sachs visited the NIH animal center in Poolesville, Maryland. He told the manager that he was interested in a nonprimate that was roughly human-size. Cows and horses were obviously too big, as were ordinary farm pigs, which can exceed half a ton when fully grown; even if an organ were harvested when the animal was young, the ramifications of placing a growing organ into a grown person were unknown.

"Have you heard of the miniature swine?" the manager said.

Sachs had not.

Miniature swine, the manager explained, were exactly what the name implied: small pigs descended from larger animals that had escaped from settlers who brought them to the New World. Said Sachs, "They got wild and they went off into the hills, into the mountains, where frequently during the year you don't have much food. And so natural selection led to the smaller pigs—the ones that could survive periods with less food. They became miniature on their own; it wasn't by intentional breeding."

Sachs liked the miniature swine for other reasons. They did not reach reproductive age as soon as mice (a quality that made the rodents attractive to researchers), but they got there much sooner than a cow or a horse, sexual maturity at five months of age. Females were fertile every three weeks, they carried their young for only 114 days, and they gave birth to litters of four to ten piglets. Furthermore, pig hearts, kidneys, lungs, intestines, and other systems were physiologically similar to human organs.

Disease-wise, pigs also had appeal. Pigs could harbor a number of pathogens that could be transmitted to people, but a carefully controlled environment could reduce that threat to an acceptable risk. No great pig-to-person disease risk had surfaced in centuries of domestication—nothing like herpesvirus simiae (B virus), which resided harmlessly in certain monkeys but was often fatal when transmitted to a person, for example, through a bite. (PERV, which would later cause alarm, had yet to be discovered.)

And there was the matter of ethics. People ate pig meat in great quantities. Who could object to pigs being used to save human lives? This was before the founding of People for the Ethical Treatment of Animals and other high-profile animal rights groups.

Having decided on miniature swine, Sachs began to inbreed three distinct strains of the animals, each of which would have the same genes within the major histocompatibility complex, the genetic region that plays a major role in rejection. In terms of transplantation, all members of a strain would essentially be identical twins. This would allow for "clean" transplant experiments that would not have been possible between randomly selected pigs whose genetic makeup was undetermined.

Using his inbred pigs, Sachs could determine the effects of antigen matching within the MHC on the outcome of transplantation—tissue matching, as it's called. He could also determine if a new drug delayed rejection more reliably than if unrelated pigs were used. Say, for example, that a kidney transplanted from one random pig into another without any immunosuppression is rejected anywhere from one to three weeks (each animal's unique genes determining the length of time) and that a transplant with similarly noninbred pigs using a new immunosuppressive drug or protocol results in rejection anywhere from a week and a half to a month. Was the drug or protocol effective? For the thirty-day survivor, yes. But what of pigs rejecting during the overlap period? A scientist could not reliably know.

With Sachs's pigs, a kidney transplanted from a member of family A into a member of family B would always be rejected at, say, seven days. If a new drug extended survival of the organ to fourteen days, then it showed promise. You could repeat the experiment, and you would achieve the same results. With "outbred" pigs, repeating an experiment would be a crapshoot, since every ordinary pig had a different genetic makeup in the region controlling rejection.

The miniature swine were suited for studies of transplants between members of the same species, and many of Sachs's advances with tolerance came out of research with these pigs. He kept his interest in xeno, but in the 1970s and 1980s his work in that area was limited to transplants of bone marrow and skin between rats and mice. Sachs wanted to transplant pig kidneys and hearts into monkeys or baboons, as others already were doing—David Cooper, for example, in South Africa and then in Oklahoma—but hyperacute rejection of large organs remained an obstacle. Although others including Cooper had somehow managed the trick, Sachs judged the cost of large-animal experiments to be prohibitive for him, a researcher who at that time was relying solely on government grants.

"That's where I was always headed—but it is such a major undertaking that it wasn't until I had industrial support from a company that it was feasible to do it," Sachs said. "It's so expensive."

When Sachs left NIH in 1991 for Harvard and Mass General Hospital, he began his collaboration with BioTransplant. Novartis would become one of the major investors in BioTransplant.

Sachs cautioned BioTransplant CEO Lebowitz that large-animal work would be expensive. "That's what we're here for," Lebowitz said. "We want to get this moving."

Sachs was ready. By now, his inbred pig herd numbered over four hundred animals.

Monkey Men

On Wednesday, March 19, 2003, one month after Goldie's goodbye, Sachs's surgeons transplanted the organs from a second double-knockout pig into a new round of baboons. The pig was numbered but not named; Goldie would be the first and only double-knockout to have that honor. The second pig had been cloned at Infigen, the Wisconsin biotechnology firm that also had a contract with Immerge.

At large-animal rounds that Friday, Sachs's scientists reported on the Wednesday operations. They had all gone smoothly, hyperacute rejection had not occurred in any animal, and all of the baboons were recuperating nicely, the researchers said.

Sachs wanted veterinarian Mike Duggan's opinion. Duggan's primary concern was the well-being of animals, not the outcome of experiments.

"Mike, are you happy with these animals?" Sachs said.

"I think they're doing extremely well," Duggan said. One baboon from Wednesday was unusually quiet, he noted, but that probably was the effect of a painkiller.

Then it was time for the weekly update on Goldie's

baboons. Thirty days after the first double-knockout transplants, the news was encouraging.

The day before, Kaz had brought B113 back to the operating room to biopsy the kidney that had been transplanted with a lobe of thymus from Goldie. Unlike the transplanted heart, this kidney was providing function: Kaz had removed one of the baboon's native kidneys (replacing it with Goldie's) and then tied off the baboon's other kidney. If the experiment failed, Kaz could untie it and the baboon would survive—a candidate for further study or retirement to a primate center somewhere.

"Kidney is pink. Looks good," said one of Kaz's assistants. Measurements of kidney function showed that Goldie's organ was performing as well as a native one. There was no sign of rejection, and B113 overall was healthy.

"This is the first time we've seen a pig kidney looking beautiful in a baboon this far out!" said Sachs. He was nursing a cold but he could barely contain his excitement. B113 seemed headed toward tolerance.

Although Sachs had begun experimenting with bone marrow as a way to induce freedom from immunosuppressive drugs and their host of side effects in the early 1980s (he worked first with mice, then with pigs and monkeys), he did not start looking into the possibilities of the thymus until the mid-1990s, after scientists elsewhere had demonstrated its potential. Kaz was one of his new research fellows then, and the Japanese doctor became enchanted with the gland.

A wonder of the immune system takes place inside it: immature T cells are transformed into germ killers.

Like all white blood cells, T cells begin as stem cells found in bone marrow. The cells migrate to the thymus, where they mature—where they are "educated" to recognize and destroy foreign or "nonself" cells. T cells are on the front line of defense against bacteria, viruses, and other forms of infection. But in the case of transplanted organs and tissues, they are a recipient's enemy.

Experimenting initially in pig-to-pig transplants and later in the pig-to-baboon model, Sachs and Kaz showed that replacing a recipient animal's thymus with the donor's thymus could help achieve tolerance.

They developed this protocol: several days before a transplant, the recipient animal's thymus was removed and the T cells circulating in its bloodstream were depleted by drugs. Kaz specialized in kidneys, and on transplant day, he transplanted the donor kidney and donor thymus into the thymectomized, T cell–depleted recipient animal. New T cells developed in the animal but matured in the transplanted pig thymus, which educated them to accept the transplanted kidney as the recipient's own. A similar mechanism was involved in the bone marrow–induced tolerance that Sachs and his colleagues had achieved with Janet McCourt, the Massachusetts woman who remained drug-free years after getting a life-saving new kidney (although in McCourt's case, tolerance was achieved with donor cells that migrated to her thymus rather than by a transplanted donor thymus).

Kaz and Sachs first reported results in 1997 in the *Journal of Experimental Medicine*. Many more papers followed.

The March 19 large-animal rounds continued.

One of Cooper's assistants gave the report on B214, the animal with Goldie's heart. "Beating well," said the researcher. "Shows no rejection."

But Cooper had elected not to biopsy the organ. A biopsy entailed putting the animal under and opening it up—a full-fledged operation with all of the accompanying risks.

"I don't blame you," Sachs said, "but it would be nice to see what it looks like."

"Let's get a few more weeks under us," Cooper said. Sachs agreed.

Sachs's fine spirits continued through the morning. During a discussion of new equipment they were buying, he noted that they would need fans for the pigs in the summer, since the animals didn't tolerate heat well. When it came to the baboons' needs, Sachs said, "Better films for the VCR? DVDs so they can jump around?" The scientists in the corridor roared.

Rounds ended at 9:00 A.M., when everyone moved to the conference room, where, almost every week, one of the fellows gave a presentation on the status of his or her work.

That morning, a Japanese fellow was speaking on her experiment involving transplanting parts of a pig pancreas into a pig kidney.

But before she began, one of Sachs's assistants noted that the system controlling the doors had failed and the doors had all locked, which necessitated using shoes and wastebaskets to keep them open.

"Do you think this has something to do with al Qaeda?" Sachs said, to more laughter. That prompted him to tell the story of the first time he had visited behind the Iron Curtain, where the checkpoints and security on leaving were extreme. "I said, 'Very interesting, a country where they don't want to let you out!'" Once again, laughter.

After the meeting, Sachs returned to his office, where he talked about his excitement. Rarely in his career had he made a point of asking how the animals had fared overnight, but he did so now every morning, worried that he would hear bad news. Sachs was not a superstitious man, but he joked that even talking about his fear would somehow bring out the "evil eye."

While nothing definitive had been proved at this early stage and much work remained, he said, "I'm more encouraged than I've ever been before."

He was so encouraged, in fact, that he wanted to ramp up the experiments, with a fresh group of double-knockout pigs that had been produced at Infigen. Having room for only sixteen baboons at his animal facility, he was hunting around the Boston area for another place to house more.

*

At one time, most baboons used in American research were caught in Africa. But private firms in Texas, Florida, and elsewhere began to breed the animals successfully. Lately Sachs had been buying his from the Mannheimer Foundation in southern Florida. They cost about $3,000 apiece, plus $500 for tests to certify that they were healthy and disease free.

The foundation had one of the more curious histories in the annals of animal breeding. Founder Hans Mannheimer, a German who arrived in New York virtually penniless in the 1930s, was first a researcher—winning several patents in chemistry, even though he held no advanced degree. When he sold his formula for baby shampoo to Johnson & Johnson, he became a wealthy man.

Mannheimer was an eccentric—a man who lived with his mother until he was forty, a man with few friends, according to his brother Walter, who described him in a newspaper story as "a genius and totally crazy." Hans slept in a reclining chair and often took dinner at midnight. He kept sea otters, tropical birds, cats, and dogs at his New Jersey residence. The sea otters had their own swimming pool, equipped with a slide.

But Mannheimer's passion was monkeys.

He owned two hundred of them, keeping some in outdoor cages and others in rooms in his house that he heated and humidified to replicate the tropics. Mannheimer hired a staff to care for his monkeys and toilet-train his four chimps. He drove his primates around in a Volkswagen bus—dressing them in snowsuits with fur-lined hoods when the weather was cold. He bought a twenty-four-foot boat, which he named the *African Queen,* and took his monkeys, dressed in little sailor suits and lifejackets, for rides on the Toms River. "Mannheimer treated those monkeys like they were his kids," a former employee said.

The crazed chemist eventually moved his pets to a fifty-six-acre facility in rural Homestead, Florida, where the climate was better suited to the animals. He created and endowed the Mannheimer Primatological Foundation with the apparent purpose of improving the care of nonhuman primates, but the language establishing the foundation was vague. After Mannheimer died in 1973, trustees began to allow surgical research and sales of monkeys and baboons.

One trustee dissented but was overruled. "Hans Mannheimer treated his animals with love, gentleness and caring," Leslie Sinclair wrote to another trustee, "and when monkeys are sold for experimental purposes, it is against the spirit and purpose of the foundation. I have been physically ill since this information came to me."

By the time of Goldie, the foundation had nearly two thousand baboons and monkeys, which the staff kept in a complex of cages and low buildings surrounded by a chain-link fence topped with barbed wire. Visitors were not welcome, but neighbors got an education of sorts in August 1992, when Hurricane Andrew struck and some fifteen hundred of the foundation's animals escaped. Most took sanctuary on roofs and in trees near the foundation, but some climbed through broken windows and wandered around inside houses, and there was a report that one baboon hurdled a fence to get at a neighbor's dog. Rumors circulated that the primates had AIDS, and many neighbors and even some authorities, believed it.

"I flagged down a police car and asked, 'What am I supposed to be doing with the monkeys at my house?'" said one man, according to a newspaper account.

"Shoot the SOBs," the policeman answered.

The neighbor did. Other citizens, cops, and even National Guardsmen did likewise, and some two hundred of Mannheimer's baboons and monkeys were killed. Said an official at the University of Miami, which also lost animals from its nonhuman primate center, "They were making target practice out of them."

From the Mannheimer center, baboons were shipped to a Veterans Administration Hospital in the Boston area that had an animal research facility where the animals were quarantined and tested to confirm that they were disease free. Baboons could harbor many germs, but tuberculosis, which could wipe out a colony, was a particular concern. Not only the baboons were tested; anyone who had contact with them had to be periodically checked for TB too.

From the VA hospital, the baboons went to Sachs's animal facility, with its levels of security. Entry to the baboon room was gained by passing through a door that opened with a code that only authorized members of Sachs's staff knew. A sign on the door reminded anyone entering of the need to wear protective clothing and eyewear. Pathogens can penetrate mucous membranes, and no one wanted contact with a baboon's bodily fluids.

The door opened into a work area, where meals were prepared, medications were stored, and baboons were readied for the operating room.

The experimental baboons occupied the larger of the two main living quarters: it was a warm, well-lit, clean, windowless area with ten or so cages. Some animals were awaiting surgery, and others had already been operated on; all wore jackets that secured the lines for administration of medications and drawing of blood samples. The animals passed the time sleeping, clinging to their perches, playing with tennis balls or rubber toys, or watching TV or wildlife videos from the small video library. *People of the Forest: The Chimps of Gombe* was the favorite. "They really do look at that one because it's very similar to them," said Willard Simmons, an attendant.

A smaller room was home to the lab's donors: a small number of baboons that provided most of the blood used in transfusions during operations. Unlike the experimental animals, the donors were named. Among them was Mongo, who had lived at Sachs's facility for nine years, the longest of all, Coolio, and Silly, so named because of the funny way his lower lip curled down. At eighty pounds, Sampson was the biggest baboon—and the dominant one, as his frequent cage rattling demonstrated. "He has to let everyone know, 'I am the biggest!'" Simmons said. Donor baboons were chosen on the basis of antibody status, weight, and proven absence of several viruses. They could donate blood no more than once a week, and every few months they were sent to the VA Hospital animal center for "vacation," where they could rest from the blood-donating duties.

Although baboons had all the appeal of the Yahoos in *Gulliver's Travels,* attendants treated them fondly. "I think it's important to talk to them—they don't have much contact with the other animals," said Meaghan Sheils. "By spending time with them, you know their personalities and you can tell when something's wrong."

Sometimes Simmons moved the cages together, allowing them to touch each other's fingers; when they did, they would chatter as well. "They have their own language," Simmons said.

Fruits and vitamin-enriched biscuits composed the bulk of the baboons' diet. But the attendants treated the animals to sugar cane and peanuts, and they constructed foraging boards: cardboard smeared with honey or peanut butter and covered with sunflower seeds or Cheerios.

Another treat was popcorn. Occasionally the attendant would pop a bowl for the baboons as they watched their videos. "They would have a blast," Simmons said.

Fresh Air

In her own closed world, a cage without bars, Cheryl Snow waited. Late winter and early spring were unkind to her.

Her health declined to the point that getting from her bed to her bedside chair was an effort. Sometimes she threw up; sometimes her blood pressure dropped to dangerous levels, leaving her faint, her eyeballs hurting. She had trouble sleeping. Pills were prescribed, but they were ineffective. When finally she drifted off, it was into a state of dreamless slumber.

Time after time, when a heart became available, it either went to someone else or final testing revealed it was unsuitable for transplantation.

The weekend of March 22 was particularly cruel.

On that Saturday morning, a nurse gave Snow the news that a heart had been found and would become hers that afternoon. The staff began preoperative preparations. Snow was a call away from the operating room when word came that the donor had hepatitis.

The next day, the staff told Snow another heart had become available. "You're kidding!" Snow said.

But a short while later, word came that the arteries to the organ were blocked. "Okay, I don't want that one, either," Snow said.

A month later, two hearts became available—but testing ruled them out too. The doctor had debated not telling Snow, but in the end he did: "I just don't want you to think the river's run dry," he explained. Snow knew that; during her long hospitalization, a fellow resident of Revere had received a new heart and gone home. Someone brought Snow a clip of his story in the local paper.

"Finding a new heart happened at the perfect time," the man, about Snow's age, told the reporter. "It was getting close. I don't know how much longer I could have gone." The man disclosed his secret to hanging on. "When you're facing death, you can never give up," he said.

Death was Snow's frequent companion during March and April. It seemed to rear up wherever she looked. On March 5, a girder crushed a construction worker who was helping build an ambulatory care center at Mass General; Snow did not see him die, but she could see the construction site from her window. Less than a month later, a woman shot and killed a staff doctor, then turned the gun on herself. "I was petrified—they had the hospital in lockdown," Snow said. "We thought somebody was running around the hospital with a gun."

And death was with Snow on her ward, a cardiac intensive care unit where transplant candidates were in the minority. Enough people died in the room next to hers that she began to call it the "death room."

"It's awful," she said. "Everybody that goes in there dies."

It didn't end there. Snow pointed to another room nearby. "Over here, a couple of people have died. And then one died down here on the right and I can't see down the hall so I don't know what the hell's going on there. I don't need this. I've got to get out of here. I can't watch any more of this." Corpses being wheeled away was not a sight anyone wanted to see.

One weekend, Snow cried repeatedly. The crying raised

her heart rate to 165, just shy of the threshold for the defibrillator connected to her heart. Medication brought the rate down, but Snow had come close to being shocked—and just imagining it unsettled her.

"They said it hurts like hell—like getting kicked in the chest by a horse. I don't want to find out, you know?"

Sachs could relate to the tribulation of being a patient—from his own experience and his father's.

Elliot Sachs was visiting his son in 1993 when a major artery in his thigh became clogged. He required an operation, which was performed at Mass General Hospital. He was discharged to the care of his wife and David and Kristina, who had designed their house with a first-story bedroom that an elderly relative might someday use.

Sachs's father never recovered. His kidneys failed and he went on dialysis, which he hated and wanted to stop. David convinced him not to, but his father was not pleased at how his life was ending. "He knew there was no hope," Sachs said. "He was not getting better. He lost all his dignity and he was such a proud man. He had to be carried to the toilet and things like that." As the end neared, most doctors would have said that Elliot belonged in the hospital, but he refused to be admitted. He died at home with his wife, son, and family around him.

"He hated hospitals," Sachs said.

He hated them for reasons that Sachs could understand

from his own experiences as a patient, foremost, of course, the memory of his childhood stay at Manhattan's Hospital for the Ruptured and Crippled. And there was the more recent memory of late summer 2002, when Sachs had been a patient at his own Mass General for an operation to cure diverticulitis.

"They gave me a super-duper room in the nicest part of the hospital—a beautiful room—and I couldn't wait to get out of there!" Sachs said. "So I do understand my father, although he took it to an extreme. I would always go in if I needed it to live, but he'd just basically had enough. He did not want to die at a hospital."

Somehow, on some days, Snow managed to lift her own spirits.

"I have to be here," she said. "I have to take care of my kids. They're the most important thing to me."

Although Snow was not supposed to have contact with the outside world, on one unseasonably warm day her cardiologist gave her permission to visit the cafeteria and go outdoors. "Just don't tell anyone!" the doctor said.

A nurse settled her into a wheelchair and hung her medications and monitors from a pole and brought her downstairs to the cafeteria, where every Wednesday Sachs had breakfast after patient rounds. Just smelling the food— which was prepared right before your eyes, not in the steamy vastness of a hospital kitchen—thrilled Snow. "You never realize what everything smells like until you haven't been there," Snow said. "Just going to the cafeteria was a big treat. I was in food heaven!"

Snow ordered two scoops of strawberry ice cream, her favorite, and the nurse wheeled her into a courtyard, where she sat in the sun for forty-five minutes eating, talking, and taking in the simple pleasures of the outdoors. "I really began to notice a lot of things we never pay attention to," she said, "the birds, squirrels, trees, the clouds."

When it was time to return, Snow said, "I'm not going back in."

But she did.

Thymo Obsessed

The spring of 2003 came and B113 flourished. Some of the medications the animal had been receiving in the immediate postoperative period were being tapered off. The hope was that it would achieve tolerance.

At rounds on Friday, April 18, researchers noted that fifty-eight days had passed since baboon 113 received Goldie's kidney. "This is the longest we've ever had a functioning xeno kidney!" Sachs said. He was enthused.

But the update on B214 was less rosy.

Cooper reported that Goldie's heart was still beating well—if you looked closely, you could see it—but the baboon didn't seem itself. Five days before, Cooper had noticed that the animal was walking with an unsteady gait, wasn't eating well, and didn't respond to eye contact, which was not like a healthy baboon at all. B214 was receiving a multitude of medications, and Cooper theorized that the animal might be experiencing a drug reaction. So he adjusted the meds, to no significant effect.

"He certainly doesn't look quite right," Cooper said. He wondered aloud if they should give steroids.

"Do what you think is right," Sachs said. His only other advice was to get a biopsy of Goldie's heart—soon. He did not want B214 to die—especially not before a biopsy, which would provide valuable clues into what had been going on inside the animal for two months.

Later that Friday, one of Cooper's assistants could find no pulse on B214's abdomen; first thing the next morning, Cooper opened the animal and confirmed that Goldie's heart had indeed stopped beating. It was enlarged and unnaturally firm, and mottled with diseased tissue, a sign of rejection. The hyperacute variety hadn't claimed it, but a more delayed type apparently had. Cooper cut the lifeless organ out and sewed B214 back up. The animal survived and recovered.

"Very disappointed," Sachs said. After two months, he'd started to believe that Goldie's heart might beat for many more.

April neared its end, bringing another disappointment, this time with B113, the xeno kidney record holder: the animal contracted an infection, apparently through a contaminated intravenous line. Despite sterile technique and good care, such contaminations occurred—it was much more difficult to have a baboon than a human as a patient, as Sachs sometimes remarked. People were not constantly jumping around, and they were considerably more sanitary than the cleanest baboon.

Antibiotics did not clear B113's infection, so on Monday, April 28, Kaz took the animal into the operating room to replace the line.

As anesthesia was being administered, the animal died.

But the kidney was fine—no signs of rejection or compromised function. It had lasted twice as long as the best of the pre-double-knockout transplants, and far longer than the control experiments transplanting double-knockout kidneys without the thymus.

Xeno tolerance seemed within reach.

"I am excited and delighted with the results so far," Sachs said. "I'm not yet convinced—but very encouraged—that this will make the kind of difference that I had hoped."

But after the experience with B214, Cooper was not so optimistic. "Very mixed results so far," he said. "Not clear-cut one way or the other."

Born in 1959 to a businessman and a housewife, Dr. Kazuhiko Yamada decided on medicine as a teenager. He did not want to follow in his father's footsteps ("I don't think it's a good thing to be in same field," he said) but he did want to make a nice living, and medicine all but guaranteed that. The deciding factor was not money, however. It was an accident that happened when he was seventeen: he was hit by a car as he was riding a motorbike in his native Tokyo. Kaz's fingers were broken, many of his teeth were knocked out, his face was scarred, and he was left unconscious for several minutes.

"It was very big accident," Kaz said. He spent a month in a hospital recovering. "That influenced me. A doctor saved my life."

Kaz earned a medical degree and a doctorate from Nippon Medical School in Tokyo, then trained in pathology, cardiovascular surgery, and urology, including transplantation. By watching other surgeons and trying his own techniques, Kaz developed an approach that emphasized innovation and speed. In a written passage, he described his style:

"Being able to operate quickly requires mastery of certain fundamentals. First is the proper way to use surgical tools, such as forceps, scissors, needle holders. Second is the ability to create a clear field to work. This sounds easy, but actually requires a great deal of skill and experience. Part of creating and maintaining a clear field is stopping any bleeding.

"A third skill is the ability to have a mental picture of the surgery. This is a particular specialty of mine. I approach every surgery with a clear vision of what the field should look like at any given moment and also what it might look like in the case of different anomalies. This allows me to operate very quickly. It has also enables me to be an innovative surgeon. For example, I discovered and pioneered a technique for transplanting vascularized thymic lobes, which others said was impossible, by picturing possible procedures in my mind.

"Finally, a surgeon must be able to handle mental excitement during surgery. If a surgeon is slow, even if they are accurate, it is because they lack one of these basic skills. These things must be learned within the first five years. By mastering all of these fundamental skills, I have become a fast surgeon."

Steady hands were another asset, and Kaz lacked the ability at first. "Actually my hands shook in the first two years because I was so concerned about looking good in front of senior doctors," he wrote. "Everybody has a weakness, and the important thing is how to overcome that weakness. In my case, I found a way to stop shaking simply by lightly touching an instrument with one finger of my empty hand." Also useful, he said, was thinking of how he dined in Japan. "I do not shake my fingers when I eat," he said. "So I thought chopsticks."

In Japan, Kaz had both a clinical practice and a research program. He served his fellowship with Sachs who asked him to stay. Accepting Sachs's offer meant giving up surgery on people—on the belief that with Sachs, he could make lasting contributions to xeno and allotransplantation. Even though he now operated only on pigs and baboons, he still considered himself a master surgeon.

"People are very surprised I am staying in research field," he said. "My wife is surprised! In all of surgery—I should not say—but I still think I am up with the best."

Kaz was named an assistant professor of surgery at the Harvard Medical School in 2000. At about the same time Goldie was born, Kaz was offered a prominent position in his native land: directorship of a large transplant program, a professorship, and a clinical practice. Kaz declined for two reasons: the possibilities that double-knockout pigs offered and Sachs.

"I respect two or three surgeons in Japan, but I can tell you my mentor is only David Sachs," Kaz said. Sachs, he said, gave him latitude.

"To achieve this goal—to find out the best regimen to induce tolerance—will be very hard," Kaz said. "It's trial and error, trial and error. Which frustrate me—it also frustrates other people. I believe it is possible—but I need a big boss who believes the same thing. Otherwise, 'Oh why you need to use such immunosuppression, why don't you decrease the immunosuppression, you have to have a longer survior.' Then we will not be able to achieve overall goal. This is very important."

The month of June neared and new double-knockout experiments began with more cloned pigs and fresh baboons. Sachs's scientists were refining their protocols: varying the dosages and duration of familiar drugs, substituting new drugs for old, using or not using pretransplant radiation, trying a different procedure with the thymus, the organ that captivated Kaz.

Rather than transplant a lobe of thymus at the same time as the kidney, the surgeon created inside the donor animal an organ unseen in nature.

Several weeks before a baboon transplant, Kaz would open a double-knockout pig, remove a lobe of its thymus, and implant it next to the pig's kidney, where new blood vessels would grow to sustain it. The resulting "thymo-kidney" would later be transplanted into a baboon whose own thymus Kaz had previously removed and whose T cells had been depleted. Thymo-kidneys seemed superior to separate thymus and kidney transplants, since the thymus part of the composite organ went into a baboon with an already-established blood supply. Kaz had seen separately transplanted thymuses shrivel and disappear when their new host failed to develop vessels to sustain them.

Having lost B113 when its kidney was still so good, Sachs took a special interest in B117 and B118, baboons that had received thymo-kidneys on the last day of April. Both did well immediately after their transplants—but then, on May 16, B117 mysteriously died. The animal had been observed coughing before being found dead, and researchers hypothesized that it had aspirated a peanut. But no peanut was found on necropsy, and the cause would remain undetermined.

"It's another tragic loss—the kidney was looking good,"

Sachs said. "But this doesn't discourage me or set me back.
It makes me sad."

With B113 and B117 gone, Sachs took an almost obsessive
interest in B118. Every morning he asked Koji Yazawa, one of
Kaz's assistants, for the animal's creatinine level, a measure
of kidney function obtained by analyzing a serum sample.
He had never before monitored an animal so closely.

At rounds on June 6, Koji Yazawa presented B118. Yazawa
said that thirty-seven days after the transplant, the baboon's
creatinine level remained stable at about .85, a healthy reading.

"This animal is okay," Kaz said. "Very active, no problem
at all."

Kaz outlined the course ahead: tapering off some of the
medications, with the hope that after several weeks they
would achieve tolerance. Kaz named some of the drugs.
They included steroids and mycophenolate mofetil (MMF),
an immunosuppressive.

But Sachs suspected the animal was receiving other
drugs too, and the fact that Kaz had neglected to mention
them made him uncharacteristically upset. "What else is
there?" Sachs said. "I want to know everything he's on."

Yazawa named more drugs.

"No heparin?"

Yes, Yazawa said; the baboon was being given heparin, a
blood thinner.

"What else?"

But the list was now complete, and Sachs was himself
again.

"This animal, knock on wood," he said, "looks very healthy, has no signs of infection. That's just great, Koji. I'm very happy."

But he wasn't happy with everything in his research, as the regular 9:00 A.M. conference on June 6 suggested.

In addition to presenting the formal results of their current experiments, scientists used the Friday morning session to pitch new ones. Today Kaz was proposing a new use for the thymus. He wanted to begin work on a creation he ultimately called the composite thymo-intestine, or "tint."

Relying on the same immunological principles involved with the thymo-kidney, Kaz sought to first conduct pig-into-pig tests—and then, assuming positive results, he wanted to transplant thymo-intestines from double-knockout pigs into baboons. The ultimate goal, as with the thymo-kidney, was tolerance.

Kaz said he would use small intestine and the clinical aim would be to cure patients with short-gut syndrome, a birth defect (or unavoidable aftermath of certain life-saving surgeries) in which insufficient bowel exists for proper nutrition. Kaz noted that human-to-human intestinal transplants remained rare due to "many unresolved obstacles." And he was right: transplanting a heart, the very essence of life, was much easier than transplanting an unsung organ that processed food.

Kaz was accustomed to receiving a warm response to his innovative ideas, especially from Sachs, but he received no applause this morning. When Sachs opened the floor to questions, research fellow Doug Johnston, who was pursuing open-heart surgery, immediately objected.

Why start with a thymo-intestine, which had never been attempted? Johnston wanted to know. Why not start with small intestine by itself?

Cooper was skeptical of any such work. "Intestinal transplantation has had much worse results clinically than most other organs," he said. Why chase a more distant dream when so much remained unresolved in their current research? "We certainly haven't answered all of the questions with kidney and heart—and for that matter, spleen."

And further, Cooper said, citing the statistic that Kaz himself had included in his proposal, short-gut syndrome affected just one to two people per million population—an infinitesimally small, if needy, group compared to the legions of people needing a new kidney, heart, liver, pancreas, or lung. "It's such a low profile," Cooper said. "I should think we would not get into it at this stage. We can't solve the problems of the other organs."

Sachs took Cooper's side.

"I'm really not sure I want to get into that," he said. "Let's think about this some more, Kaz. It sounds like we need to do some more thinking."

Kaz had been pacing, his face increasingly furrowed, his arms folded in growing disgust. Meekness had not made him a Harvard professor of surgery.

"So—no thymo-intestine!" he said. "I'm very surprised. No study [of mine] has been rejected!"

Kaz shut off the overhead projector and gathered up his slides. He was in a huff.

Sachs tried to calm him. He was not refusing to allow the experiments, he said, he just wanted more information. "The reason for these meetings is to get just this kind of input," Sachs said.

The surgeons in the room peppered Kaz with questions about the technical aspects of how Kaz proposed to marry thymus to kidney. The Japanese doctor was quickly on the defensive. He wasn't used to being grilled.

"What is it you want to do?" Sachs finally said.

Kaz gave an answer that satisfied no one.

It was almost ten o'clock, and Sachs had had enough. "I have to leave," he said. "Right now. I have to run."

Sachs walked out of the room.

Kaz and Johnston exchanged a few more words, and then Cooper left too. The meeting was over. Kaz had not won approval, nor would he. But his preoccupation with thymo organs had become a bit of a running joke with some of Sachs's researchers. "People have been saying the next thing we should do is the thymo-penis, to facilitate sex-change operations," said one scientist.

Money Matters

If tempers were uncharacteristically short that June 6, Sachs, Kaz, and Cooper could be forgiven. They had more than science on their minds. They were worried about money.

They were hardly the first scientists to be so concerned. Decades ago, some researchers used their own money or relied on the largesse of friends and relatives to fund their work. Others relied on universities and philanthropists. And there was modest assistance from the federal government, some states, and certain private corporations. The revenue stream was a patchwork quilt that varied year to year, project to project, never with any guarantee that a researcher's needs would all be met.

The period after World War II saw the federal government become an increasingly important source of funding for medical research, especially from the National Institutes of Health, which had a budget of $2.8 million in 1945 and $27 billion in 2003. Some of that funding has gone into staffing and development of NIH centers, but most has gone to scientists in the form of grants. The agency awarded $18.4 billion in research grants (46,081 separate grants) in 2003, more than double the dollar amount (and almost double the total number of grants) compared to just a decade before. Over the years, Sachs had received tens of millions of dollars for his many experiments, and NIH funding remained the primary source of revenue for his transplantation biology research center. According to federal guidelines, grants were awarded based on merit, experience, and need, among other factors. Final decisions were made by committees in a process similar to decisions on publication in peer-review journals.

Companies in the medical science industry traditionally conducted their research in-house: they could directly control it, they could conveniently bring together science and marketing, and, with competitors always interested to discover what the next great money maker would be, they could better maintain secrecy. But that self-reliance began to change in the 1970s, said Thomas F. Boat, M.D., director of the Cincinnati Children's Research Foundation at Cincinnati Children's Hospital Medical Center. More and more corporations learned that outsourcing work could produce cost savings and some hospital and university centers had capabilities that they lacked in-house; academic researchers, typically strapped for resources, signed on in growing numbers. "Over the last two or three decades, that has been slowly mounting," Boat said. "In the last ten years, there's been a lot more of it. There's been a paradigm shift."

Corporations in partnership with Mass General Hospital scientists agree that the hospital will own the patent for successful development of a technology, treatment, or product; in return, the hospital agrees to license the patent to the company for commercialization. The royalty is negotiable, according to Sachs, but is generally less than 10 percent of the net profits from the innovation. The hospital gets the lion's share of this royalty income, with "some small fraction" going to the scientist's lab and "an even smaller fraction to the scientist," Sachs says.

Some corporations demand secrecy agreements forbidding a scientist from discussing research with a colleague before they will fund an outsider. Sachs will not sign such a document. But that could have a downside, as Sachs noted in a letter to a conflict of interest committee he served on: "Many companies refuse to enter into a collaboration with an academic scientist without a written agreement that all results will be kept secret for a specified period of time. In the most common situation, they will not provide their product or assistance without such an agreement, despite the knowledge that both parties will lose the benefit that might arise from the collaboration. The situation is particularly sad when one considers the potential loss of research which might lead to a cure for a disease. However, I agree with my institution's policy that it is unreasonable for an academic scientist not to be allowed to discuss his or her research with scientific colleagues." Sachs proposed a solution: "If scientists at all major academic institutions were to adopt a common policy of not accepting industrial agreements that limit their rights to discuss their data with their scientific colleagues, the problem would disappear. I believe that companies would realize that their needs were better served by agreeing to permit such scientific interactions and have the advantage of academic collaborations than to continue to limit them and suffer the loss of this important source of discovery and advice." No such policy has been universally adopted.

Every academic scientist spends considerable time seeking funding for his or her work. As application deadlines approach for substantial NIH grants, for example, an entire lab might stop virtually all of its scientific work so that fellows, staff members, and the director can complete the paperwork, which often runs into hundreds of pages.

And it never ends.

"I don't think that anybody who tries to do original work has an easy time of money," said Dr. Joseph P. Vacanti, the world's most prominent tissue engineer, who is a colleague of Sachs's at Harvard Medical School and Mass General. One of Vacanti's most promising projects is an engineered liver, an early prototype of which has functioned successfully in rats.

"Getting government to give you a grant is not simple," Vacanti says. "Getting industry to give you a grant is not simple. And getting somebody to contribute a buck just from the goodness of his heart is not easy. There's not a day that goes by that, in one way or another, I'm not working on the money part of it, whether it's talking to philanthropists, talking to company people, writing a government grant, talking to the government. It's a continuous, ongoing part of what you do, like brushing your teeth."

Such is the lot of the academic scientist, and it drives some of them from academia to the private sector. But those who stay do so because they can remain independent researchers and still practice medicine, like Vacanti, who still performs liver transplants, or keep a connection to patients, like Sachs.

"I really enjoy clinical medicine and I enjoy research. Although it's hard to get the grants sometimes, [academia] is the only place you can do both of those unfettered," says Dr. Peter Quesenberry, an adult stem cell scientist at Roger Williams Medical Center in Providence, Rhode Island, and a professor at Boston University School of Medicine.

"If you go into a corporate environment, you're going to be restricted. What usually happens in that environment is people love it there the first couple of years—they've got a lot of resources and money, a beautiful office, their own parking space, all that crap. Then what happens is in the third or fourth year or so the suits will come in and say, 'Well, we think you should maybe go this other direction. Or now we want you to focus just on this: the development of a money-making drug.'"

High-risk projects such as xenotransplantation draw a cautious response from the NIH; while Sachs had won grants for his work in the field, even with his reputation he had never succeeded in getting the level of funding he needed from the federal government alone to move as fast as he wanted. That was why Novartis's support of Immerge was so critical to him.

And that was the real reason he and some of his researchers were so ornery that June 6. Word had recently reached Boston that executives at the Novartis headquarters in Basel, Switzerland, were strongly inclined to stop funding

Immerge and xenotransplantation. A final decision would be made over the summer.

The loss to Sachs would be over a million dollars a year—about 10 percent of the annual budget for his center. Heaven only knew where he could make up such a shortfall.

CHAPTER 4

Risky Business

Goat Glands

Since ancient times, animals have benefited people. They have protected, warmed, labored, herded, transported, and provided companionship. Animals have been used for food, drink, and clothing. It was not much of a leap for the relationship to expand to include the more intimate one of donor.

Although attempts likely took place earlier, the first documented animal-to-human transplant was in Padua, Italy, in 1628, when sheep blood was transfused into a man, apparently in hopes of curing some disease. The notion caught on: sheep blood transfusions were soon reported in London, and the practice continued well into the nineteenth century. Treatment was not limited to physical ailments. Acting on the assumption that a sheep's docile nature might be transferred through its blood, doctors gave transfusions to the violently mentally ill. It was a beguiling theory, but nothing more. Sheep blood had no beneficial effect on people.

Like sheep-blood transfusions, tissue transplants became something of a rage in the 1800s, with lamb skin and frog skin used to treat ulcers and burns—often with success but not because a person's immune system accepted lamb or frog matter. Rejection prevented the foreign skin from grafting, but it served as a protective covering while the human tissues below healed themselves.

The early twentieth century brought the first attempts to transplant solid animal organs into people—which, from a technical point of view alone, was a more complicated undertaking than simply applying a piece of frog to an area of injured skin. The immunology was the same, however, and the attempts were all doomed.

In 1902 a Viennese doctor trying to cure renal disease sewed a pig kidney onto a woman's arm, where he had good access to vessels lying close to the surface, and three years later, a French physician put pieces of rabbit kidney into the kidney of a dying child with the same intent. Neither experiment succeeded, nor did an attempt in 1906 to transplant a goat kidney into a person. The early age of organ xenotransplantation ended in 1909 with the work of Berlin doctor Ernst Unger, who started unconventionally: he transplanted a kidney from a stillborn human baby into a baboon. The baboon promptly died. Unger then went in the other direction, transplanting an ape kidney into a woman. That failed too.

Most scientists saw little immediate promise in animal-to-people transplants. Legitimate transplant researchers were concentrating on finding a way to get the human body to accept tissues and organs from another person. They were interested in allotransplantation.

But xenotransplantation did not disappear. In the 1920s, thanks in part to a colorful American quack and a dashing Russian émigré who lived in Paris, xeno received unprecedented publicity. It was not the sort of attention any scientist dedicated to unraveling immunological puzzles appreciated, but it did, in a perverse way, inform the public of a potential new avenue for medicine.

The quack was John R. Brinkley, and he grew up in the mountains of North Carolina, where superstition and the old folk ways still held influence a century ago. Brinkley attended medical school but never graduated—yet he called himself a doctor, and with a diploma purchased from Eclectic Medical University of Kansas City, Missouri, he was able to practice medicine in states, including Arkansas and Kansas, that recognized the document.

Moving from town to town with his bag of instruments, Brinkley performed tonsillectomies and appendectomies and operated on hernias, often in patients' homes. In 1917 he opened a drugstore in the small town of Milford, Kansas. One day a farmer dropped by and asked to speak privately with him. The farmer said that he and his wife had a son and wanted another, but the farmer had become impotent.

"No pep," he confided. "A flat tire."

Brinkley told the farmer that he knew of attempts to cure impotency through various drugs and even jolts of electrical current, but none, to his knowledge, worked. Then Brinkley recalled an earlier job he had held at a meatpacking plant, where federal inspectors had told him that the goat meat best suited for human consumption came from animals with the healthiest testicles. "You wouldn't have any trouble if you had a pair of those buck glands in you," Brinkley said, apparently in jest.

The farmer took him seriously. "Well, why don't you put 'em in?" he said.

"It might kill you," Brinkley said.

"You go ahead," the farmer said. "I'll take the risk."

"What if people found out?"

"Who's to hear about it unless you tell them?"

"I don't have a goat," Brinkley said.

But the farmer did. "I raise 'em," he said. "I'll furnish the goat. You do the operation."

Brinkley's biographers do not record the results of his first transplant, except to say the farmer was satisfied and paid Brinkley $150, a handsome sum for the time—and would have paid more if he had been able. But they do tell the story of the second transplant, into a man by the name of William Stittsworth, who, like the farmer, had been unable to give his wife a child. A month after operating on the man, Brinkley operated on the woman, transplanting a goat ovary into her. A year after that, the biographers assert, the couple had a boy, whom they named Billy. It was not the only ludicrous touch surrounding the man who was called the goat gland doctor. Brinkley wore a goatee.

Word of Brinkley's operations spread, and patients abounded. Secretive at first, Brinkley began to boast about his work, which, he claimed, not only reversed impotency but also rejuvenated the whole person, male or female. Brinkley coined a catchy slogan, "You are only as old as your glands," and he made his claims in advertisements and published rebuttals to legitimate doctors and medical associations, including the American Medical Association, which accused him of fraud and worked to get his various licenses revoked, eventually with success. Brinkley opened a hospital and raised his fee to $750–$2,000 on the rare occasions when the gold standard of his form of rejuvenation, a condemned human prisoner's testicles, became available. It was an early echo of the organ trade in China, which human rights activists would condemn decades later.

News accounts were filled with praise for Brinkley—and rife with exaggeration, with one reporter claiming Brinkley was the only surgeon who had ever "successfully transplanted an eye from one living creature to another." Harry Chandler, owner of the *Los Angeles Times,* brought Brinkley to California in 1922 to operate on him and members of his staff and then published flattering stories of the results in his influential newspaper. Brinkley also operated on several movie stars, reportedly earning $40,000 for his West Coast trip. Brinkley became so well known in America that one of the popular jokes of the 1920s went: "What's the fastest thing on four legs? A goat passing Dr. Brinkley's hospital."

The truth of Brinkley's surgery seemed to be that, in some cases, when impotence had a psychological basis, the operations worked as a placebo: a man was emboldened to perform by believing the miraculous doctor had cured him. Brinkley, of course, did not trumpet his failures—although they were publicized in later lawsuits. He never published scientific accounts of his estimated sixteen thousand operations or the short- or long-term results, and he never fully explained the procedure, beyond saying that he transplanted slices of a goat testis into a human testis and furnished the foreign tissue with a blood supply. Brinkley had obviously overcome no immunological hurdle, and his implants all must have been rejected. Some of the grafts became infected, a situation that was unlikely to cure impotence. Others atrophied inside their host—becoming small, inert, and useless.

All Brinkley said about the principles involved in his "successes" during one of many legal proceedings against him was that the slices of goat gland "were gradually absorbed" by a patient's body. "I don't mean to say that the little thing lived," Brinkley said. "If it became infected . . . it would spoil."

Other gland doctors surfaced, but none achieved Brinkley's fame in America. In Europe, the celebrated gland doctor was the Russian-born Parisian, Dr. Sergei Voronoff, a surgeon who by 1926 had performed about a thousand transplants.

Like Brinkley, Voronoff promised rejuvenation—his operations, he claimed, could turn the clock back by up to thirty years. More amazing still, he claimed, people he treated could pass life-extending benefits on to their children. And while it is doubtful he ever actually experimented on a child, he even alleged that his transplants could make an already gifted young person smarter. "I call for children of genius," he said. "Give me such children and I will create a new super-race of men of genius." This was a Frankenstenian spin on eugenics, a popular movement of the time.

Unlike Brinkley, Voronoff used monkey and baboon testicles in his operations (and occasionally human testicles). Quack though he was, he had a better inkling of immunology than Brinkley did.

"To use the glands of other animals is to ignore completely the laws of biology," Voronoff wrote. "They could never be, in the human organism, anything but foreign bodies." Part showman, part madman, Voronoff at one point kept hundreds of monkeys imported from Africa in a palazzo on the Italian Riviera. He courted the press and was rewarded with flattering accounts of his brilliance. He was honored by the French and Spanish governments. He lectured at medical conferences and was praised in the pages of *Scientific American* in 1926, at the height of his popularity: "Even death, save by accident, may become unknown if the daring experiments of Dr. Sergei Voronoff, brilliant French surgeon, continue to produce results such as have startled the world."

In his book, *Rejuvenation by Grafting* (1925), Voronoff wrote of the amazing transformation of a seventy-four-year-old man who "walked with difficulty, leaning on a stick. . . . His memory was bad, his intelligence slow and sluggish. He presented all the characteristics of the senile type. He had been sexually impotent for 12 years." Eight months after Voronoff operated, according to the surgeon, "The man was literally 15–20 years younger. His whole condition, physical, mental, and sexual, had undergone a radical change. The grafting had transformed a senile, impotent, pitiful old being into a vigorous man, in full possession of his faculties."

Until he was discredited in the 1930s by researchers who could not repeat his alleged results in animals, Voronoff was held by some to be a pioneer.

An Awesome Fungus

In 1954 researchers finally figured out how to successfully transplant an organ from one person into another. That year, doctors at Boston's Brigham and Women's Hospital transplanted a kidney from a man into his identical twin—an accomplishment for which surgeon Joseph E. Murray and immunologist E. Donnall Thomas received the 1990 Nobel Prize.

Their success encouraged a generation of doctors, including Starzl and Barnard. But the early going in person-to-person transplants was dicey, with high mortality rates and diminished life expectancies. Rejection remained a hurdle as scientists sought to understand the complex immune mechanisms involved and to develop the immunosuppressive drugs and treatments that would curb the body's natural defense against foreign invaders.

The early 1980s introduction of cyclosporine, the most effective immunosuppressive drug to that point, brought allotransplantation from the realm of semiexperimental medicine to established practice. Stories of successful organ transplants, which had been a lock for page 1, began to move to the inside sections of the daily newspapers.

Cyclosporine was developed by the pharmaceutical firm Sandoz, which later merged with Ciba-Geigy to become Novartis. It may never have existed if not for an offbeat but foresighted Sandoz tradition. Starting in 1957, employees who were traveling on work or vacation were encouraged to take soil samples wherever they went; the employees placed the samples in plastic bags and brought them back to the laboratory, where they were analyzed for the presence of fungi that might be used for new antibiotics. Analyzing soil samples from Norway and Wisconsin, Sandoz scientists in 1970 isolated a fungus they had never seen: *Tolypocladium inflatum*. Metabolites derived from the fungus proved to have no antibacterial potential, but further tests revealed it was the most powerful immunosuppressive Sandoz researchers had ever seen. Animal trials were conducted in the mid-1970s, and the first human tests began in late 1976. In 1983 the U.S. Food and Drug Administration approved the drug for use in preventing rejection.

Cyclosporine was a true wonder drug: rejection rates dropped, life expectancies rose, and, encouraged by such positive outcomes, more and more medical centers began to offer transplants. Waiting lists grew as, for the first time since the pioneering work of Murray, Starzl, and Barnard, the demand for human organs in the United States began to outstrip the supply.

Cyclosporine generated millions in profits for Sandoz every year (and still does for Novartis, under the brand name Sandimmune), but the growing waiting lists limited the market potential: more transplants were needed to sell more of the drug. Executives observed that American public awareness campaigns weren't solving the shortage, and they knew artificial organs wouldn't either, at least for the foreseeable future. And stem cell therapy, tissue engineering, and more exotic technologies such as nanomedicine were still years away.

So Sandoz began to look at animals.

And what they found wasn't limited to a handful of crude early-twentieth-century experiments and the glandular quackery of Brinkley and Voronoff. Some of the allotransplant pioneers, along with other scientists, had dabbled in xeno. They learned of Starzl's six (failed) attempts to put baboon kidneys into people in 1964, three years before his landmark human liver operation. Even the greatest transplant pioneer of all, Barnard, had tried to put a baboon heart and a chimpanzee heart into people in 1977. Like the others, he had failed.

There were more. In 1965 Mississippi's James Hardy transplanted the first chimpanzee heart into a person, a sixty-eight-year-old comatose man who died after two hours; Texan Denton Cooley, the famous open-heart surgeon, transplanted a sheep heart into a person in 1968; and English surgeon Ross, one of Cooper's early influences, twice tried to transplant a pig heart into a person. Cooley and Ross were both foiled by hyperacute rejection: even before the patients had been sewn up, their immune systems had left the animal organs a lifeless black mess.

And there was Tulane University's Reemtsma, who managed to keep a woman alive for nine months with chimpanzee kidneys. The public would never accept chimps as organ donors, but Reemtsma at least had demonstrated that xenotransplantation was more than quackery.

No xeno case achieved more fame than that of Baby Fae, the pseudonym given to the twelve-day-old California girl who was born in 1984 with a heart defect so severe that only a transplant could save her.

With human donor organs that size exceedingly scarce—it took a rare tragedy to leave an infant brain-dead—Loma

Linda University Medical Center surgeon Leonard L. Bailey decided to give the baby a baboon heart. An accomplished and respected cardiac surgeon, Bailey had performed animal-to-animal experiments, but he had never tried to put an animal organ into a person. He tried this time because he believed Baby Fae had no other option, and the parents, desperate to save their child, let her be a guinea pig.

The operation took place on October 26, 1984.

Almost before leaving the recovery room, Baby Fae, photographed but never identified (at the request of her parents), became a global sensation. Hundreds of journalists with their notebooks and tape recorders and cameras and remote-broadcast vans descended on Loma Linda; the whole world, it seemed, waited expectantly for the latest news.

And the news, the first few days, seemed scripted from a fairy tale: the girl survived the operation and her body did not reject the baboon heart.

But the news quickly turned. The infant's kidneys began to fail, and soon her heart was ebbing too. Bailey and his team placed Baby Fae in an oxygen tent and ran her blood through a dialysis machine, but they could not save her. On November 15, less than three weeks after the transplant, she died.

In some of his statements, Bailey was optimistic. "Baby Fae has opened new vistas for all, including the as-yet-unborn infants with similar lethal heart disease," he told the press. "The Baby Faes and their parents are the real pioneers in this quest to enrich our quality of life. Her unique place in our memories will derive from what she and her parents have done to give rise to a ray of hope for the babies to come."

But the experience left Bailey rattled. A media mob had descended on his hospital, and animal rights activists had protested by waving signs that said "Stop the Torture" and drawing comparisons of the surgeon to Joseph Mengele, the infamous Nazi doctor. Bailey didn't see any such comparison, of course, but philosophically he was conflicted. Although his hospital's ethics committee had given him permission to transplant more baboon hearts into people, he chose not to. He chose instead to concentrate on human-to-human transplants as he sorted through the many ethical issues Baby Fae brought to the forefront of public debate.

Speaking a decade after the operation to the author of *Knife to the Heart,* which chronicled the case, Bailey said, "Some say that an animal's life is just as valuable as the human baby's life and that we have no business taking the animal's life. I'm sensitive to that point of view; I have some misgivings about it myself. But as I look around the planet earth and the various species that are scattered about, there is a certain survival of the fittest that goes on in nature, the lion eats the zebra and so forth, and it seems to me, as a human being, that my obligation is to my own species first.

"If there is a way to support the life of a member of my own species, even though it requires the sensitive and caring taking of a subhuman primate life, I think I'm obligated to pursue that course of action . . . [but] it gives you pause for reflection. You have to continually check what you are doing against public opinion, whether or not it is justifiable, whether it is right."

Sandoz committed resources to xeno in 1991, the year Sachs returned to Harvard. Its first major investments were in Imutran and BioTransplant. While Sachs, collaborating with BioTransplant, experimented with bone marrow as a means of inducing tolerance in pig-to-baboon transplants, Imutran's chief scientist, Dr. David J.G. White, developed a line of transgenic pigs that, coupled with immunosuppressive drugs, provided a degree of resistance to hyperacute rejection. White's pigs had such potential that Sachs, who had spent many years developing his own line, experimented with some of them.

For a spell, White and his pigs were the toast of xenotransplantation. And not just insiders raised the cup: writers from the mainstream press treated him as a celebrity—the scientist almost singlehandedly working to save many who were otherwise doomed.

"The nice thing about the pig is you know who the donor is going to be weeks or months before the transplant," White told the *Saturday Evening Post* in 1998. "It's not like you're waiting for someone to fall off his bicycle." Asked by the *Post* reporter when people might benefit, White said, "I would be very disappointed if we didn't start our transplantation program before the end of the millennium." The writer stated, "Here he exudes confidence!"

In a subsequent interview with the *Post*, White addressed the possibility that a pig organ could transmit a disease: "Most people would rather be alive than dead," he said. "If you say that you can be dead in three months or you can have a pig-heart transplant with a risk that in twenty years you might get leukemia, there are no choices. It's an option other than death."

If there was a golden age of xenotransplantation, the mid- to late 1990s was it. Sandoz was in the game, as were two other large firms, Baxter and Geron. Several smaller startups went into the business. Universities sponsored programs. The science advanced and commercial prospects seemed to improve. BioTransplant CEO Lebowitz told the *Boston Globe* that he expected to be using pig organs in people by the year 2001. "I think we will see pig hearts transplanted in primates within the next twelve months, but using pig organs in people is still about five years out," Lebowitz said. BioTransplant had just made its first public stock offering, raising almost $25 million. Things were close to giddy.

The premier issue of *Xenotransplantation*, the first scientific journal devoted exclusively to the field, was published in August 1994. The editorial board included Cooper, Benedict Cosimi, Starzl, David White, and Megan Sykes, one of Sachs's principal scientists and a prominent bone marrow transplant researcher. Sachs was the journal's editor in chief.

Starzl contributed an article on his two failed attempts to transplant baboon livers into people in 1992 and 1993. Nonetheless he expressed a degree of optimism that "the xenograft barrier may be more vulnerable than most people realize at present" and xenotransplantation could become a reality. Cooper and his colleagues in Oklahoma City wrote about the role of the galactose sugar molecule in hyperacute rejection. Mass General infectious disease specialist Jay A. Fishman wrote about strategies to prevent transmission of pathogens from donor animals to patients.

In his opening editorial, "Xenotransplantation: Challenge and Opportunity," Sachs posed the fundamental question and gave an unequivocal answer. "Will xenotransplants work?" Sachs wrote. "I believe the answer is yes. I state this because 1) none of the many obstacles to success I can envision appears to be insurmountable; and 2) given the present shortage of donor organs, the transplant surgeon is obliged to find another donor species." Sachs went on to predict that pigs, not nonhuman primates, would likely be the donors. And he argued that the best avenue to long-term acceptance was tolerance, which, he noted, was being researched at several centers, including his own Transplantation Biology Research Center.

Xenotransplantation seemed within reach and Cooper, writing with Lanza in a 1997 article for *Scientific American,* conjured up the wonder of the dawning day. "Early morning, sometime in the near future," the scientists wrote. "A team of surgeons removes the heart, lungs, liver, kidneys and pancreas from a donor, whereupon a medical technician packs these organs in ice and rushes them to a nearby airport. A few hours later, the heart and liver land in one city, the two kidneys in another, and the lungs and pancreas arrive in a third. Speedily conveyed to hospitals in each city, these organs are transplanted into patients who are desperately ill.

"The replacements function well, and six people received a new lease on life. Back at the donor center, surgeons repeat the procedure several times, and additional transplants take place at a score of facilities distributed around the country. In all, surgical teams scattered throughout the U.S. conduct more than 100 transplant operations on this day alone.

"How could so many donor organs have possibly been found? Easily—by obtaining organs not from human cadavers but from pigs. Although such a medical miracle is not yet possible, we and other researchers are taking definite steps toward it."

But as the twenty-first century neared, xeno began to lose some of its luster, at least to the biggest corporate backer.

Imutran's scientists managed to keep White's pig kidneys functioning inside baboons for as long as seventy-eight days, but not all had normal function. The firm's average survival for hearts was only thirty-nine days. (Cooper managed to get one to 139 days, but excluding that case, his average with a series of ten was only twenty-six days. Six in the series died before a month, and two lived just a week or less.) Imutran asked the U.S. Food and Drug Administration and several transplant surgeons what threshold in pig-to-baboon experiments they would want to see before sanctioning trials on people. The consensus was six months of functional survival in more than 50 percent of all pig-to-baboon experiments. Imutran was not even close.

"We did not achieve this criterion," said Paul Herrling, head of corporate research for Novartis. "It turned out things were more complicated than assumed."

Hot Zones

PERV is one of the viruses that over the centuries have attached themselves to the genes of many life forms without causing harm to the host. There was no evidence that PERV had ever crossed over into man in vivo: pigs and people had coexisted for centuries and the growth of the pork industry brought thousands of workers into daily contact with millions of pigs that ended their lives in bloody slaughter, but there was no known case of PERV infecting a person. "If PERV could infect humans, it would have done so by now," wrote the Islet Foundation, a group that supported research into pig pancreatic islet transplants as a cure for diabetes, one of Immerge's other research interests. Pancreatic islets produce insulin.

Tests in the 1990s, however, revealed that the virus could enter certain human cell lines in the lab. Novartis found itself involved in an unwelcome controversy. Activists and scientists, including even some xeno researchers, raised an alarm. Virologists especially were concerned: they knew well how unpredictable a virus could be.

Although PERV had never been known to infect (or activate inside) a person, who could guarantee it never would? If it did, what sort of disease would it cause? Could that disease spread into the population? If so, could it be stopped? Or would it be like AIDS, which defied total containment and cure after one of the most intensive public health campaigns in history?

And who could say that there weren't other pig viruses, as yet undiscovered, hiding on the genome?

"This danger of creating a new disease was not only a risk affecting only the patient her- or himself—it was all of a sudden creating a [possible] new threat to society," said Herrling. Herrling believed that the FDA and other regulatory bodies could eventually be convinced that the risk was sufficiently remote that clinical trials could proceed. But with progress slow on other fronts at Imutran, Novartis was softening on the company.

Although it was not the crucial factor in the decision to close Imutran, a highly publicized animal rights controversy developed, centered around Novartis—not the sort of thing to warm the heart of a company officer who must answer to stockholders.

The controversy arose in the fall of 1998, when the BBC broadcast a report by two British animal rights groups, Compassion in World Farming and the British Union for the Abolition of Vivisection. The groups claimed that White's work represented "infectious risks to patients and the public," that mingling an animal's immune system with a person's could have unforeseen consequences, that there was "scant evidence" that animal organs could sustain a human life. If they were able to, wrote neurochemist and zoologist Dr. Gill Langley, one of the authors of the report, a patient living with one would no longer be truly human. "It's now clear that a human xenotransplant patient will become a literal chimera, a pig or baboon human hybrid," wrote Langley. "Not only will an animal organ produce animal factors which will circulate around the blood stream, but cells from the animal organ will travel all over the human body and to every organ, including the heart, lungs, kidneys, bone marrow and lymph nodes."

White defended his work, asserting that the shortage of organs made xeno research imperative—and that his company was overseen by a government agency. "We are responsible to the Xenotransplantation Interim Regulatory Authority which looks at all these ethical issues as well as the issue of safety and efficacy, and we're very happy to abide by them," White told the BBC. The Compassion in

World Farming and the British Union for the Abolition of Vivisection controversy passed, but a worse one loomed on the horizon.

It broke in September 2000. Based on leaks of Imutran data to Uncaged Campaigns, a U.K. animal rights group, the "Diaries of Despair," published in newspapers and online, painted a picture of mad scientists who abused animals in pursuit of "savage science." "Behind the smokescreen of confidentiality exists a culture of contempt for the rule of law and for the welfare of animals. This is a tragic scandal of historic proportions."

The "Diaries of Despair" included detailed accounts of surgery on the baboons and monkeys at Imutran that received White's pig hearts and kidneys and day-by-day observations of the postoperative course of the animals that survived. The accounts were gory, with little indication that the primates had been treated for suffering and pain (although Imutran insisted they were). Exaggerated or not, it was a damning report, and it concluded with a condemnation of the entire field of xenotransplantation:

"In five years of experiments, causing severe suffering to hundreds of primates and sacrificing thousands of pigs, Imutran and Novartis managed to squeeze average survival times up by a couple of weeks. Very little progress was made in overcoming the profound immunological obstacles to xenotransplantation. Is the notion of functioning pig organ transplants nothing more than scientific arrogance driven by personal ambition and commercial greed? Transforming organs from pigs into human organs appears to be nothing more than a modern form of alchemy."

A month after the controversy broke, Novartis decided to close Imutran—a coincidence, the company said. The animal rights activists believed otherwise. To support their belief, they noted the fate of White, once the toast of xeno. He left the country to join Canada's John P. Robarts Research Institute and the University of Western Ontario. They were fine institutions both, but they weren't Harvard.

Although he could not speak to what happened at Imutran, Sachs knew of animals that had been treated cruelly elsewhere.

"There's always an element of truth—there are studies that have been done that were unnecessary," Sachs said. "There have been abuses in the past and I think that's what's lent credence to some of this movement."

Sachs aimed to treat his animals like human patients whenever possible. "We have the privilege of being able to work with animals," he said. "It's a privilege that we don't want to abuse. I believe in humane treatment of animals but not in animal rights."

Sachs acknowledged that a sick patient who needs surgery is unlike a healthy animal that doesn't: a sick person seeks to get well, while a healthy animal, if it had the choice, would prefer not to be subjected to risks of injury or death unrelated to its own well-being. "That's a big difference," Sachs said. "And when I interview anybody who wants to work here I try to assure that they understand this difference. I point out that these animals are very well treated, just like patients, but they don't need the operation—they come in perfectly healthy, not needing a transplant."

The Sachs lab was approved by the Association for Assessment and Accreditation of Laboratory Animal Care, whose members included the American Medical Association and the American Veterinary Medical Association. All Mass General Hospital animal facilities were subject to internal regulation as well as regulation by the U.S. Department of Agriculture, which periodically conducted surprise inspections. All experiments had to be approved by the Mass General Subcommittee on Research Animal Care before they started. Such outside oversight and internal control were relatively new. It wasn't so long ago that researchers using live animals answered only to themselves.

Mass General's research animal subcommittee was composed of thirty-seven members having authority to deny an application or stop an experiment in process. In 2004, twenty-five members were scientists from the departments of medicine, surgery, radiation oncology, radiology, pediatrics, anesthesia, critical care, pathology, and oromaxilofacial surgery. Five members were veterinarians. Three belonged to the hospital's safety office, which governed use of biological, radioactive, and chemical agents. A hospital senior vice president held a seat. Three members came from the community: a lawyer, a teacher, and a homemaker who once worked in animal research administration.

The subcommittee met regularly in the main conference room of Lawrence House, headquarters for the hospital's trustees. It was an oval room with floor-to-ceiling windows, a brass chandelier, and, over the fireplace, an oil painting of former trustee, millionaire John R. Macomber with two of his hunting dogs. An American flag stood on one side of the fireplace, while a Mass General Hospital flag—there really was such a thing—stood on the other. The atmosphere was very Bostonian, very Harvard.

During one meeting, subcommittee members reviewed a proposal by Dr. Joren C. Madsen, a cardiothoracic surgeon and one of Sach's senior investigators. Like Cooper, Madsen had an interest in heart xenotransplantation, but his research at the moment mainly involved transplanting hearts and lungs between the same species: pig to pig, baboon to baboon.

Madsen proposed experiments that Kaz would follow carefully: transplanting hearts with thymus. Madsen wanted to use baboons. The heart and thymus would be harvested from a baboon, leaving the animal dead; another baboon would receive the the heart as a secondary, or heterotopic, organ. The proposal was thirty-three pages long, most of them responses to questions on the subcommittee's standard application form. Anyone proposing an animal experiment completed this form and then faced the subcommittee's scrutiny—in written critiques and then in person at one or more meetings.

"The scientific reliance on live animals should be minimized," the form stated. "Alternative models such as mathematical, computer simulation, or in vitro biological systems can sometimes be used to replace animals. Explain why the use of animals is necessary for this experiment."

Madsen (and his fellows) wrote, in an answer that would have applied to most of the work in Sachs's lab, "The complexity of the immune mechanisms under study make computer and in vitro models complementary at best."

What is the significance of this study?

"Cardiac transplantation has become an effective treatment modality for patients with end-stage heart disease. However, this dramatic improvement in early patient survival has uncovered a late and lethal complication of heart transplantation—accelerated allograft coronary arteriosclerosis. Accelerated allograft coronary arteriosclerosis affects up to 60 percent of all cardiac allograft recipients and is a major limitation to the long-term survival of heart transplant patients." Use of the thymus, Madsen wrote, might also help achieve the Holy Grail.

Why is this species appropriate?

Madsen had checked three boxes: "The process resembles that in humans," "Tissues or other substances to be harvested require an animal of this size," and "The size or anatomy of this species is best or uniquely suited to the procedure to be done." Pigs could not be used, Madsen wrote, because "significant differences exist in the anatomic arrangement of the thymus between swine and humans and other primates, making baboons an ideal model."

What were the expected results?

"Graft loss due to acute or chronic rejection will be the major endpoint of the study. We believe that cotransplantation of heart and thymus will lead to prolonged graft survival, and maybe lead to tolerance."

Minimizing suffering was a required objective. Under the USDA animal-use pain categories, Madsen checked off category D: "Teaching, research experiments or tests conducted involve PAIN or DISTRESS, and for which appropriate anesthetic, analgesic or tranquilizing drugs will be used."

Are there any alternatives that would involve less pain?

"The literature search conducted indicates that there are no alternative procedures that would involve less pain or distress."

What is the plan for posttransplant care?

"After any surgical procedure, the baboon is monitored carefully by the surgical team and veterinarian for any signs of pain or distress. Additional analgesics are given as needed. If pain or distress cannot be eliminated, the animal will be euthanized according to AVMA guidelines."

When the experiments are completed, what will happen to the baboons?

"All baboons in this study will be maintained for long-term survival. If the transplanted heart or en-bloc heart thymus graft is rejected, every effort will be made to maintain the baboon for further study. When a heterotopic heart is rejected, it will be removed under general anesthesia. Surviving baboons will be followed up for the development of alloantibody and in vitro assays.

"In those instances in which no reasonable expectation of survival can be entertained, or in which the scientific objectives of the experiment have been achieved, euthanasia will be performed by IV injection of a lethal dose of sodium pentobarbital. This method of euthanasia is consistent with the recommendations of the Panel on Euthanasia of the American Veterinary Medical Association. In any case that the animal's tissues are not needed to complete the studies and the animal is considered 'healthy' by the veterinarian, efforts will be made to retire the animal to a breeding colony, such as that at the Southwest Foundation in Texas."

The subcommittee had the power to reject an application, return it for revision, or approve it as submitted. The members approved Madsen's application with little debate, and the study began some while later. The early results supported Madsen's hypothesis: use of the thymus extended survival over hearts transplanted alone.

"Eventually we intend to try it for xeno," Sachs said, "but we don't have funding for that yet." Xeno research was never cheap.

The Rights of Species

Sachs, of course, placed the rights of people above the rights of animals.

A poster hanging in his animal facility reminded his researchers of that philosophy. Distributed by the Foundation for Biomedical Research, a nonprofit group chaired by pioneering heart surgeon Dr. Michael E. DeBakey, the poster showed a group of English animal rights protestors and was captioned, "Thanks to animal research, they'll be able to protest 20.8 years longer. According to the U.S. Department of Health and Human Services, animal research has helped extend our life expectancy by 20.8 years. Of course, how you choose to spend those extra years is up to you."

Sachs told the story of speaking in the late 1980s at his oldest daughter's high school, which had an animal rights club.

"I started off by asking the class: How many of you know anyone who had polio? Not a single hand went up. Then I told them what polio was like when I was their age. I explained that the only way that we would ever have been able to cure that was through animal research."

One boy asked why Sachs didn't just use computers.

Because if scientists don't fully understand the immune system, the immunologist replied, how could they program the computers?

"So I went on and on and we talked about my research and how important it is to have humane treatment of animals that need to be used to cure disease." Sachs laughed. "About a week later Michelle came home and told me that they had disbanded the animal rights club."

Sachs particularly objected to People for the Ethical Treatment of Animals, the group that, according to its Web site, www.peta.org, "believes that animals deserve the most basic rights—consideration of their own best interests regardless of whether they are useful to humans. Like you, they are capable of suffering and have interests in leading their own lives; therefore, they are not ours to use—for food, clothing, entertainment, or experimentation, or for any other reason."

Troy Seidle, PETA's director of science policy, maintains that xenotransplantation is wrong for reasons beyond the group's basic philosophy "of one group in power taking advantage of another group that isn't in power," as he puts it. "Xeno adds a number of additional kinks to it," he said. "We already know that viruses can jump from animals to people. What happens when you put a pig organ in an immunosuppressed person? That's the public health side."

Another side was the argument xeno proponents made about filling a shortage many described as urgent. Seidle rejected the assertion: he maintained that American society could improve on its human organ donation rates to close the gap. "Have we really made every effort to increase donations?" he said. He noted that America has an "opt-in" policy, by which a potential donor must declare his intention to donate. But other countries, including some in Europe, have an "opt-out" policy, in which all people are assumed to be donors unless they specify they do not wish to. "Given the alternatives, which have yet to be fully explored; the high risks in terms of public health concern; and the animal welfare concerns on top of that, we feel it grossly premature for government regulators to even be considering cross-species transplantation." Seidle is also troubled by the spiritual implications of a person who would be part—conceivably many parts—pig. "How do we define what it is to be human? What are the implications of having a genetically engineered organ in your body? There are some very serious ethical issues that people should stop and look at."

Another group, the New York–based Campaign for Responsible Transplantation, an anti-xeno group that counts Greenpeace and the Jane Goodall Institute among its members, finds an irony in using pig organs to save some people whose consumption of pork and other meats can be a factor in heart disease and other ailments. The group's home page features a composite photograph of a man with a pig snout.

"Scientific studies have demonstrated that pigs are highly intelligent and sensitive animals. Pigs used in studies at the University of Pennsylvania manipulated joysticks with their mouths to solve mazes and play games on a computer," the group says. "Policy-makers in the U.S. and elsewhere have decided that it is 'ethical' to use pigs in xenotransplants because pigs are killed for food. But two wrongs do not make a right. Ironically, it is precisely because people eat too many pigs, and have unhealthy lifestyles, that pig organ transplants are being considered. A large majority of heart, liver, and kidney transplants could be prevented if people reduced their meat, (and alcohol and tobacco consumption). We should ask whether it is acceptable to make pigs and other nonhuman animals scapegoats for our species' self-destructive behaviors."

Peter Singer, professor of bioethics at Princeton University and author of *Animal Liberation* (1975), the book that launched the modern American animal rights movement, believes that xeno is a cruel concept unlikely to ever be realized.

"We've been told that success with xenotransplantation is 'imminent' every year for about the last ten years," Singer says. "It is a huge waste of research money and animal suffering, because we could do a lot more to increase the supply of human organs, e.g. by having an 'opt out' system, as some other nations do, rather than an 'opt in' one; and by having a regulated market for organs, too; and even if this works, it is only an interim solution until we learn to grow human organs from stem cells."

The tactics of some animal rights groups bothered Sachs nearly as much as their philosophy. He maintained that some of them circulated photos of animals in surgery and then deliberately misrepresented the circumstances, alleging that the animals had not been anesthetized, for example. And truth be told, it could be difficult for a layperson viewing a photograph of a baboon on an operating table to determine what was really going on—whether the animal was painlessly asleep or consciously enduring cruel suffering.

"That's the kind of thing that gets stolen from the files and then sent out in their inflammatory literature," Sachs said. "Rather than saying this is a fully anesthetized animal who's being studied and is being taken care of and monitored just like humans are, they say this animal is being tortured. You take some widow who's inherited a large amount of money from her husband and has, let's say, no children, and loves her dogs and cats. And they send a picture like that and say, 'This is what's happening in these laboratories to cats and dogs, please send your money so we can save these animals.' That is how PETA has become so wealthy."

Although the animal rights movement had not directly affected his work, Sachs had read the accounts of research animals being freed, including some of Bailey's not long after the Baby Fae case; of laboratory windows being broken; of a laboratory being torched; of one scientist being beaten up and another hung in effigy; of leafleting and marches and acts of vandalism. These events disturbed Sachs, and he regretted not speaking out more as he had to his daughter's class in an effort to enlighten the public. The silence of most animal researchers on the topic gave the upper hand to the activists.

"In general, we don't do enough," Sachs said. "We assume that because this is a lunatic fringe that it's not going to affect us and that's wrong—it does. I don't spend enough time on it because I have too many other things I need my time for. But it would be valuable to get to the young people in this country and show the truth of what's going on, as I did with that class."

Although in the clear minority, a few researchers did what Sachs advocated. One was Bailey, who defied Loma Linda policy by allowing journalists from the *Sacramento Bee* to observe and photograph baboons in surgery; Bailey told reporter Deborah Blum, who won a Pulitzer Prize for her series on animal experimentation, that he refused to be "blackmailed" by activists. Another outspoken scientist was Dr. Stuart Zola-Morgan of the University of California at San Diego. He researches the effects of brain damage—induced by his surgery—on memory. Zola-Morgan keeps a framed certificate that reads "Vivisector of the Year" on his desk.

"The old model is gone," Zola-Morgan told Blum. "The animal rights activists are so good at what they do, if we keep pretending they'll go away, we'll lose everything."

That may have been hyperbole, but Gallup polling over the years showed declining support for the use of animals in medical experiments. A 1989 Gallup poll showed that 77 percent of Americans supported such research, but that number had fallen to 62 percent in the 2004 poll. Sixty-two percent was nonetheless a solid majority. Wrote a Gallup editor: "Just under two-thirds of Americans believe that medical testing on animals is morally acceptable, while 32 percent feel it is morally wrong. This finding is not completely surprising—medical testing can include researching cures for deadly diseases such as AIDS and cancer, and many people believe that animals' lives are worth the sacrifice."

In the ethical discussion of xeno, Sachs and those in his field found an ally in the Catholic Church, whose Pontifical Academy for Life established a task force on animal-to-people transplants. Its twenty-four-page report, sanctioned by the Vatican and published in 2001, refuted the argument that all species should enjoy equal rights.

"We do well to consider what the intention of the Creator was in bringing animals into existence," the report stated. "Since they are creatures, animals have their own specific value which man must recognize and respect. However, God placed them, together with the other nonhuman creatures, at the service of man, so that man could achieve his overall development through them." The Vatican insisted that animals be treated humanely: "unnecessary animal suffering must be prevented." And it insisted that use of animals must meet a proven human need. Xeno, it said, did.

"We reaffirm that humans have a unique and higher dignity. However, humans must also answer to the Creator for the manner in which they treat animals. As a consequence, the sacrifice of animals can be justified only if required to achieve an important benefit for man, as is the case with xenotransplantation of organs or tissues to man—even when this involves experiments on animals and/or genetically modifying them."

Sachs happened to be in Rome for a meeting of a transplant society when the Vatican issued the report. Along with other scientists, he was invited to meet the pope. Catholics were blessed and non-Catholics shook his hand. "I have a picture if you'd like to see it," Sachs said.

Sachs produced a snapshot of himself with the pope. He

laughed. "If I had known, I would have dressed properly! I had no tie or jacket—it was a hot day!"

Stressful Times

Novartis closed Imutran in 2000, but it was not ready to let xeno go.

Sachs's research and the progress at BioTransplant impressed the Swiss company, and it created a new firm, Immerge BioTherapeutics, in which it owned a 67 percent share; BioTransplant owned the rest (elements of Imutran were incorporated into the new firm). Immerge was based in Boston—across the street from Sachs's lab—and headed by Greenstein. Immerge began operations on January 1, 2001. Novartis agreed to give the new firm $10 million a year for three years. (Sachs had no stock or other financial interest in this or any other firm.)

In a press release announcing the new company, Herrling spoke enthusiastically of merging Novartis's experience with immunosuppression and Sachs's advances with tolerance and his line of inbred pigs.

"From a scientific and business perspective, this move makes sense for both companies and provides a great opportunity to maximize our complementary technologies," Herrling said. "By joining the two approaches, we hope to bring forward the day when xenotransplantation will become a clinical reality. Novartis is committed to research in xenotransplantation as part of its long-term program to deliver new solutions to the worldwide organ donor shortage."

Sachs told the publication *Nature Medicine* that he was "extremely pleased" at Immerge's founding. "People have now realized that this problem is more difficult than first thought," Sachs said. "Additional methods such as tolerance induction are needed because the amount of immunosuppressive drugs required for a xenotransplant will be so great the side effects will be prohibitive."

As they reviewed Immerge's results at the beginning of 2003, the last year of the three-year commitment, Novartis executives saw scientific progress—and this was before Sachs's experiments with the double-knockout pigs. Along with advances in tolerance, Sachs, working with Immerge, had demonstrated that his line of miniature pigs contained a form of PERV that apparently was not transmitted to human cells in culture.

But after spending almost $100 million on xeno, according to Herrling, Novartis did not see a return on its investment anytime soon. (Others, suspecting that the pharmaceutical company did not wish to disclose the full extent of its losses, estimated the true figure was much higher.)

"The results, while extremely interesting from a scientific point of view, were still not close enough to a practical application," Herrling said. "We are not a university who is doing science for science's sake—we need at the end of the day when we invest to have a product coming out of it."

Novartis had already notified Immerge in the spring of 2003 that it was likely to get out of xeno, but it had not closed the door entirely. It seemed unlikely that anything could change the company's mind, Herrling said, but a final decision would not come until late summer.

More results from the double-knockout experiments would be available then.

It was around this time when Sachs began to joke that he wished he was buddies with Bill Gates.

He still hoped that Novartis would change its mind, but in the all-but-certain event it didn't, he and Immerge needed to find new sources of funding. Sachs's allotransplantation experiments would be unaffected, but he doubted he could continue his xeno research at the present level with only government grants, even if all of his existing awards were renewed and he managed to win new ones.

At first Sachs left matters largely in the hands of Greenstein. She wrote a business plan, polished her presentation, and began courting venture capitalists. Sometimes Sachs went along to lend his name.

Greenstein herself had credentials. She held master's and doctoral degrees in microbiology, and, like Hawley, the scientist who'd knocked out the sugar gene, she had completed a postdoctoral fellowship at the Dana-Farber Cancer Institute. She had taught pathology at Harvard Medical

School, including one year as an assistant professor. After serving as vice president of ImmuLogic Pharmaceutical Corporation, a small Massachusetts biotech firm that eventually went bankrupt, Greenstein joined BioTransplant. She was senior vice president and chief scientific officer when Immerge was founded and she took charge. Greenstein was author or coauthor of fifty-four published papers. And in the hunt for investors, it didn't hurt that her husband was Dr. Paul Bleicher, an immunologist and dermatologist who left medicine to found a biotech company, Phase Forward, that had grown to 350 employees. In the biomedical and investment worlds, Bleicher had valuable connections.

And yet Greenstein's was no easy job.

The economy in general was soft in the spring of 2003, and it was not as if Immerge had created a sexy new search engine or computer; it was selling an expensive, unproved science with public health risks far beyond the usual. The Goliath in the field, Novartis, was throwing in the towel. And the era of stem cells, which could put a twinkle in an investor's eye even though the science remained in its infancy, had arrived. The actors Christopher Reeve and Michael J. Fox were publicizing that cause, but no celebrities were lending their names to genetically engineered pigs.

In her presentations to prospective investors, Greenstein described Immerge's successes with manipulating genes and cloning—the ability to knock out both copies of the sugar gene and then produce, at Infigen and in Missouri, pigs from Sachs's inbred line for experimentation.

She spoke of Sachs's successes and those of another of Immerge's partners: the University of Minnesota's Diabetes Institute for Immunology and Transplantation. Immerge and the institute were transplanting pig pancreatic islets, which make insulin, in the hope of curing diabetes. Using diabetic monkeys as their animal model, scientists had achieved functioning survival of islets for more than 180 days, a record. Dr. Bernhard Hering, the institute's associate director, said when Immerge announced the results on June 3, "This data is very significant because while we have been able to reverse diabetes in past islet studies, we had only seen two- to three-week survival times before the graft was lost due to the overwhelming rejection response. The survival times we are reporting on today should only increase as we further optimize the immunosuppressive regimens."

The islet research was so promising, Greenstein informed the money people, that Immerge hoped to begin human trials by the end of 2005. Solid-organ transplants, she predicted, could go clinical as soon as a year or so later. The double-knockouts gave her that confidence.

To save money on rent, Immerge moved from its original location near Sachs's lab to a biotechnology office park near the Massachusetts Institute of Technology in Cambridge, one subway stop from the main campus of Mass General Hospital. A collection of stuffed pink pigs adorned Greenstein's office, along with a Christmas card from PPL Therapeutics featuring a picture of the world's first cloned mammal that was captioned: "Season's Greetings from Dolly." From Greenstein's window, you could see the building that housed a branch of Novartis's U.S. operations. The pharmaceutical firm was prospering. It would end 2003 with a profit of more than $5 billion, a figure that would please company executives and stockholders.

One spring day, Greenstein sat at her desk and discussed Immerge's strategy, which was to raise $25–30 million from private investors, enough to keep Immerge going for three more years and enough, hopefully, to bring xeno to clinical trials. The reception she was getting, Greenstein said, was a cautious one. The golden age had passed.

"There have been a lot of companies that have come and gone in xenotransplantation," she said, "and they're a bit concerned about that. It's high risk and it also has high cash requirements. We get a lot of concern about the amount of money that would be needed both to get from here to the clinic and then through the clinical trial process to real revenues, which is of course what they care about."

Investors worried that animal organs had never been sold before, which meant there was no road map to follow in gaining FDA approval. PERV also troubled investors, even though Sachs's pigs did not transmit the virus to human cells in culture, as did the possibility that some other as yet undiscovered virus could be hiding on the genome. Liability for the next plague was not an enticement.

Despite these concerns, Greenstein remained optimistic.

"I would hope that with good results," she said, "that we'd be able to get enough financing to continue the work."

But the pool of potential investors for such high-risk science was even smaller now. PPL Therapeutics had sold its xeno business to the University of Pittsburgh's Starzl Institute and a small group of investors who had formed a xeno company called Revivicor. PPL's high-profile public relations campaign had finally paid off.

At 10:30 on the morning of Monday, May 12, the day after Mother's Day, a nurse came into Cheryl Snow's room in the Cardiac Intensive Care unit at Mass General Hospital. Snow had not left the hospital in almost three months. Her vigil continued interminably.

"No food or drink," the nurse said.

Snow was nonplussed; as recently as the preceding Friday, yet another heart had become available. It was not an ideal heart, the doctors said: its coronary vessels were blocked, and she would need bypass surgery using vessels cut from her legs for the transplant to work. But the heart was hers if she wanted it. She didn't.

"Why would I want my veins cut?" Snow said. "That's crazy. I would be going through enough with transplant surgery. I didn't need any extra surgery."

On May 12, Snow was prepared to be disappointed again.

But as the afternoon wore on and no bad news arrived, she began to think, *This must be it.*

Snow called her family and told them to come to the hospital. A friend washed her hair and Snow shaved her legs. They listened to an ABBA recording and they watched a tape of *Sex and the City.* They tidied the room.

At 4:00 P.M., a doctor dropped by. "It may be a go," he said. The test results were all looking good.

Snow grew anxious; a nurse gave her a sedative to calm her down, but its effect was limited. Snow's sister started crying and Snow cried too.

At 8:30 P.M., Snow was summoned to the operating room. Nurses wished her luck and her family accompanied her as she was wheeled away. At the door to the operating suite, Snow kissed everyone and said she loved them.

"See you when I wake up," she said, and she was gone.

Snow remembers the operating room being cold. She was lying awake at 11:30 P.M. when surgeons received word that final testing had confirmed the heart was good. Snow would later learn that it had come from a twenty-two-year-old, but that was all she learned. She didn't immediately ask for more details; she didn't want to know. She was troubled that someone else's life had to end for the rest of her life to begin.

Moments later, the anesthesiologist said, "We're going to put you to sleep now."

A mask covered Snow's face. The anesthesiologist told her to start counting backwards from 100. At 97, she was out.

Chief surgeon Dr. Arvind Agnihotri and assistant Dr.

Joren Madsen opened the woman's chest and inserted the lines that would connect her to the bypass machine, which would keep her alive during the transplant. When the heart courier was ten minutes away, his precious cargo chilling in ice water inside a cooler, the surgeons started the bypass and began to lower Snow's temperature. Tissues survive longer when cold.

The surgeons clamped the vessels to Snow's diseased heart and now her blood was flowing around the heart and through the machine. The old heart came out easily and the new one was brought to the table. The surgeons sewed it in without complication. Then they began to rewarm their patient. The clamps were removed and blood entered Snow's new heart.

"The heart began to beat soon thereafter," Agnihorti wrote in his operative notes. "A defibrillatory shock was not required."

As summer turned to fall, Novartis finalized its decision to leave xeno. Not even the results from Sachs's double-knock-out experiments could persuade the company to stay.

"It was a step in the right direction," Herrling said, "but in practical terms, it did not significantly change the equation." After more than a decade, Novartis was still not making money from xeno, and Herrling questioned if anyone ever could. He wondered if xeno would ever be perfected or would remain the tantalyzing dream of scientists like Sachs.

"Who knows?" Herrling said. "It's very, very difficult to predict."

Sachs could not guarantee that the puzzle would be solved either, but he believed he was closer than anyone ever had been. By October, he had a large body of data demonstrating the advantages of his double-knockout pigs.

Hyperacute rejection had been overcome. B118, the baboon with the thymo-kidney he had watched so carefully, prospered for eighty-one days, a world record—and it died of pneumonia, not the delayed form of rejection, against which the double-knockouts also seemed to provide some effectiveness. The other kidney recipients did not live that long, but most died of complications, not rejection. Most of the heart transplants survived more than two months, although none beat Cooper's record of 139 days using one of White's transgenic pigs. Still, the double-knockout hearts seemed to require less immunosuppression, which in itself was significant. No one had yet attempted a thymo-heart with a double-knockout, but Sachs wanted to. The betting was that if he did, hearts would last longer.

Sachs wanted to expand his experiments, not worry that they would be curtailed—or stopped altogether.

"Before," he said, "after twenty to thirty days we would always lose the organs. Now suddenly we're going up to sixty and eighty days with no change in our protocol other than the use of the knockout pig as a donor. And what's even more exciting is that when these animals have died, it's been from infection and the organs were still functioning and looked relatively normal on histology. This is something we've never seen before.

"My major concern right now is not whether the biology and the immunology are capable of overcoming the rest of the problems. It's whether or not we're going to be able to continue to get sufficient funding for these studies. It's more of a worry to me now than the biology, which is a very unusual place for me because it's the first time the actual results are anywhere near this encouraging—and yet it's also the first time when the funding situation has looked so critical."

Sachs was working into the evenings on grants—applications for new ones and renewals of existing ones—and he had assigned a junior member of his staff to search for philanthropists; his joke about Bill Gates was really no joke at all. He continued to work with Greenstein as she sought venture capitalists. He went to the Mass General Hospital administration, where he received assurances but no cash. He wrote to the president of Novartis but didn't received a response.

Circumstances had forced one of the world's foremost transplant immunologists to all but beg.

"I would just like to be able to spend all my time on the science," Sachs said. "It's a very stressful time."

On Monday, November 17, Sachs led his scientists through the pig rooms, where all could see and learn firsthand the condition of the animals. It was a weekly event, a comple-ment to every Friday's large-animal rounds. Cage by cage Sachs and his entourage went, stopping for an update at each pig.

One of them, a double-knockout, was asleep; it had been given a sedative and would soon be moved to the operating room.

The scientists left the pigs and headed into the hall, where a new poster had been placed on the wall. It showed a toddler staring at an assortment of cleaning products in a cabinet under a kitchen sink. "Why are household products tested on laboratory animals? Ask somebody with kids," the poster read.

After a stop in the baboon room, where three of the animals were being readied for surgery and the rest were eating breakfast, Sachs parted company with his people. "Good luck today," he said.

As the day unfolded, they would need it.

Kaz wasted no time in scrubbing, gowning, masking, and opening the double-knockout pig, which lay anesthetized on the table. Kaz had been here already, a few weeks before, when he had created a thymo-kidney inside the animal; today he would transplant the composite organ into baboon 134. In a separate operation, he would transplant a lobe of thymus with the pig's other kidney into B135, which an assistant was opening on the table across from the pig. The purpose of these parallel experiments was to help determine which approach—which use of thymus, composite organ or separate implant—was the better route to tolerance.

The adjacent OR, meanwhile, was ready to receive B229, into which Cooper would transplant the pig's heart. Dor, the Dutch research fellow, carried the sleeping baboon in and placed it on the table.

"She's very pretty," said the technician, Meaghan Sheils.

"Good biceps," said Dor.

"She's buff," said Sheils.

"Chances are this is the last knockout transplant of the year," said Dor. He was correct: months would pass before another double-knockout pig visited Sachs's OR.

Dor prepped B229 and Cooper scrubbed in. But before he took up his scalpel, Jim Winter and Mike Duggan dropped by to issue a caution. Due to some twenty transfusions given recently to save a baboon of Kaz's that had become anemic, Winter said, the blood supply was perilously low. Sachs's donor baboons would not be able to help, Duggan explained, because they were "tapped out" after giving all that blood. Blood could be purchased from the Mannheimer Institute, but it wouldn't arrive until tomorrow, and it was expensive, Winter said, about $200 per unit. With money so tight, this was not the time for premium baboon blood, if there was any way around it.

"If everything goes all right, we should be okay," Cooper said. But everything did not always go okay in surgery, as every surgeon knew.

Winter had other worries as well.

The lab was short on sterile ice, he said, so they would have to be miserly with its use. Another concern was the room ventilation, which had failed over the weekend. The faulty units had been fixed, but starting them up now, with animals' insides exposed and air blowing down from the ceiling vents, would increase the risk of infection as the dust and dirt stirred up during repairs blew through. Winter considered covering the animals with sterile drapes until the air blew clean, but no one knew how long that would take. So they decided to go without ventilation. It was the safe choice, but it carried consequences. It meant the two operating rooms would get warm.

Yet another worry was one of the anesthesia machines: a line had blown and it was still out of commission, which meant they had only three functioning operating tables. It was an inconvenience more than anything else, but in addition to the other problems, the timing was regrettable.

"It's kind of everything coming together, unfortunately," Winter lamented. If one were so inclined, one could find a metaphor for the funding issues in the aggravations of the day.

But nothing deterred the unflappable Kaz. By noon, he had removed the pig's thymus and kidney and was diligently sewing it into a baboon. The technician administered cardioplegia, the pig's heart stopped, and in minutes, Cooper had it out. He and his assistant, Yau-Lin Tseng, a research fellow from Taiwan, trimmed the organ and brought it to baboon 229.

They had nearly finished transplanting it when Cooper said, "Is it just me or is it hot in here?" It wasn't just him; in the absence of ventilation, the room had heated up. Cooper asked Sheils to wipe his brow and clean his glasses. He wondered aloud if the temperature would affect the pig heart during the time it took to establish a blood supply from the baboon.

A few minutes later, Cooper ran into more trouble: uncontrolled bleeding in the operative field, a dangerous condition because clear vision can be lost. Cooper, who'd transplanted hundreds and hundreds of human and animal hearts, was uncharacteristically flustered.

"Put this clamp on better!" he instructed Tseng. "Bring it up more! Use the smaller one! The smaller clamp!"

The bleeding persisted.

"Have you got any blood?" Cooper said to Winter.

Winter returned with a unit, one of the last in stock. Sheils transfused it, and Cooper and Tseng brought the bleeding under control.

Sachs, meanwhile, had come into the adjacent room, where Kaz was nearly finished. The Japanese surgeon explained how he had sewn the double-knockout kidney into the baboon. "Very pretty," Sachs said. "Beautiful. Remarkable."

Sachs then visited Cooper. "How's it going?" he said.

"Not too bad," Cooper said. He was calm now. They had not run out of blood or sterile ice.

"See you later," Sachs said.

"Okay."

Six minutes later, Cooper said, "Could we have the defibrillator ready?"

The heart started on the second try.

"Looking better now," Cooper said.

At large-animal rounds the following Friday morning, the scientists reported that all three of the baboons from Monday's operations were doing well. So were the baboons from the transplants performed on November 6. One of those animals was B228, recipient of a pig heart.

No one could predict, of course, that B228 would set a world record—but it did.

As pressing as money had become as the year wound down, it did not flavor everything, and the mood at rounds on that Friday was upbeat. Cooper was pleased with B223, which had supported a pig heart for one hundred days. The organ continued to beat strongly, and the immunosuppressive drugs were being tapered off. An upcoming biopsy would reveal more details about the condition of the heart, and Cooper suspected he would find some long-term damage—damage he believed they could learn to prevent. "There are other antibodies that we're not seeing," he said, "but I'm pretty optimistic we'll solve the problem in the next couple of years. We just need to plod on."

Suddenly there was a terrible banging in the ceiling. It probably was just steam pipes, but it startled everyone.

"What is that?" Sachs said.

"It's that fellow who came ten years ago, trying to get

out!" Cooper said, with dry British humor. Everyone laughed.

Rounds ended and the conference room filled. Scientists, fellows, and staff helped themselves to the ample array of doughnuts, bagels, orange juice, and coffee that always awaited. Once Krispy Kremes reached the Boston market, Dunkin' Donuts were history.

There was another item on the table: a box of cookies that a technician baked in honor of B223. The box cover had a photo of the animal in its cage. It was holding a cup of yogurt and seemed to be smiling. B223 was a lab favorite, an unusually friendly animal who liked to "talk" to the staff.

"Happy Day 100, B223!" the photo caption said.

But the celebration would prove to be short-lived.

During an examination of B223 the next week, Cooper felt a weakened pig heart in its belly: he rated its beat 2.5 on a scale of 3. A few days later, he lowered the rating to 2. On December 1, he lowered it to 1.

"It's on its way out," Cooper said.

The surgeon decided that he should examine the organ while it lived; it had now survived 110 days inside its host, longer than any other heart from a double-knockout pig. It had made history.

As Cooper and his assistant scrubbed in, the animal was sedated, shaved, and carried to the operating room. It stirred on the way.

"Go to sleep, sweetie," said Sheils. "Go to sleep."

The animal was laid on the table and anesthetized. Cooper placed his hand over where the double-knockout heart had been transplanted.

"It's beating more weakly," Cooper said. "It wouldn't support circulation now. It would have last week. The only question now is, Should we take it out, Kenji?"

Dr. Kawaki, Cooper's assistant, had no answer.

"We'll see how it looks," Cooper said. Drugs could be used to strengthen its beat, but experience had shown that if the organ was in trouble, drugs would prolong things only briefly.

Cooper and Kawaki opened the baboon's abdomen and the pig heart came into view. It was a dusky purple, not the pink of perfect health.

"Definitely going to take it out?" said Winter.

"Yes," Cooper said, "otherwise we'll be back in a couple of days. Would someone get the camera?"

A technician snapped a few shots and Kawaki clamped the vessels connecting the heart to the baboon.

"Cut above them," Cooper said. "Take the heart out."

With scalpel and scissors, Kawaki did.

Cooper held the organ in his hand for a moment. "This heart's still beating a little bit, see it?" he said. The organ was not giving up easily.

Cooper placed the heart in a bowl of saline water, where it beat four more times and stopped.

Pathologist Akira Shimizu appeared in the room with his sample jars and, using an ordinary fish knife, sliced cross-sections of the organ. One section quivered a final time and that was it. With a scissors, Shimizu snipped off several slivers of heart muscle, which he deposited in the jars.

Cooper and Kawaki were closing B223's abdomen. Shimizu packed his jars into a cardboard box, bowed, said, "Thank you," and left.

Shimizu's studies showed that the surface and vessels of the swollen heart had been damaged, that the heart had hemorrhaged, and that blood had coagulated inside it. Cooper again theorized the cause was a baboon antibody they had yet to discover. Eventually, he said, they would have to find a way to protect the heart—or knock out the gene that produced the molecule against which the antibody was reacting.

It would take time and money.

Sachs was running short on both.

The end of the year was approaching, and employees at Immerge had already been told to expect layoffs after the holidays.

Cooper, meanwhile, had been quietly approached by the Starzl Institute, the only other center to have double-knockout pigs. The Starzl group wanted Cooper to head its xeno-transplantation program. Cooper was interested. Starzl's program was well funded and Cooper was concerned about the financial future of Immerge and the impact on Sachs's xeno work.

He knew, as did Kaz and Sachs, that in the twenty-first century, high-risk research like theirs needed a Bill Gates— or support from a corporation or investors.

CHAPTER 5

THE VALUE OF A LIFE

Einstein on the Wall

In their annual holiday letter, David and Kristina Sachs and their children brought relatives and friends up to date on the year just ending. Teviah, their youngest child, wrote about being in his second year of medical school. Daughter Karin had become engaged. Daughter Jessica had been accepted to a fellowship in pediatric hematology and oncology at Boston Children's Hospital. And daughter Michelle and husband Idriz were expecting a baby—David and Kristina's first grandchild—in May.

Kristina began her remarks noting the death of the family dog. "We quickly adopted Cannon, who helped us get over this loss," she wrote. She mentioned the two-week trip through Scandinavia the family had taken during the summer, and she celebrated the news about the baby on the way. She wrote that in March, she had turned sixty.

David ended the holiday letter with a bittersweet passage:

"This year started sadly with the loss of my mother just before the New Year. I sometimes find it hard to believe that both she and my father are now gone, but I am thankful that they both reached old age and knew much joy in their lives. Kristina now often tells me that I sound just like my father—which I always take as a compliment even if that is not always her intent!"

Sachs mentioned his research and the burden that Novartis's withdrawal had placed on him: "Fortunately, our work has seen some outstanding progress this year, which should help us to achieve the additional funding we will need." He wrote of his travels—with his family and to professional meetings—and of two reunions he had attended, his thirty-fifth at Harvard Medical School and his fortieth at Harvard College. "I was surprised to note that some of my classmates appear to be getting old!" he said.

One of the photographs illustrating the holiday letter showed Kristina with Cannon, an abandoned dog they got from a German shepherd rescue league, by the fireplace in the family room. It was a large room decorated with plants and featuring an extensive video library and a widescreen TV. Kristina had designed the house, and to create an inviting space that brought people comfortably together, she kept one entire side open: no doors or walls separating family room, kitchen, dining area, and Kristina's alcove, where she kept her computer. In good weather, the doors opened to a large deck and, beyond, a green lawn and flowerbeds that Kristina tended.

David's study was on the opposite side of the house, at the end of a long hall lined with black-and-white photographs of David and Kristina as children, their parents, and some of their Old World ancestors. Annual photographs of the current Sachs family hung there too.

Sachs's study was not as bright as the family area—it had northern exposure—but it offered a view of one of Sachs's favorite places, his vegetable garden. That a fruit-bearing plant could emerge from a seed still awed him.

"I just like to see a seed come up. It comes up like that and it starts to spread its leaves. It's awe at nature."

Sachs's garden was small, barren now in winter, but "very fertile, very productive" in season, Sachs had said during a tour early that September. Showing a visitor around, he pointed out the tomatoes, which were ripening in abundance; the rhubarb, already past season; the corn; and the edamames, soybeans that are eaten from the shell. Sachs was particularly excited about his sunflowers. They were the tallest he'd ever raised.

"Aren't they incredible?" he said.

More amazing still, he said, was what had happened while he'd been on vacation: bean and cucumber vines had climbed the sunflower stalks, and many of the vegetables had ripened high in the air. For a moment, he sounded like the kid he was in Yonkers. "Can you see 'em? There have been several big ones. I'm thinking maybe I'll stake 'em next year and then have stringbeans and cucumbers hanging from my sunflowers, sort of like a pole." It would economize space in the small garden, Sachs said, and perhaps there was some additional, as yet undiscovered, advantage to having vegetables ripen in the air. His experience with the sunflowers reminded him of one of the principles that guided him in his research. "It's what Pasteur always said: 'Chance favors the prepared mind.'"

Sachs's study featured the usual computer, papers, and books, and a guitar that the immunologist occasionally played. He also had a ukulele tucked away somewhere. As in his office, portraits of people covered the walls.

One photo was of President John F. Kennedy, whose liberal politics were close to the scientist's own, and another was of his early mentor in chemistry, Louis Fieser. He had a picture of his mother and a picture of his father, who spent the last months of his life living with the Sachses. Still another photo was of his Uncle Teviah, who had influenced Sachs as a child so strongly that he named his only son after him. Teviah passed on when David was a teenager, leaving the nephew with an unsettling feeling that his uncle had taken some great secret knowledge with him to his grave.

"'David,' he used to say, 'when you're a little older there are so many things that I can tell you about life from my experiences.' I never got to know what they were." It saddened Sachs still.

Raised a Jew, Sachs loved the traditions of Judaism but didn't take the teachings literally; in grade school, his reasoning had led him to a conclusion about religions that claim to be the one true faith. "It couldn't be possible that one was right and all the others were wrong."

But Sachs was far from an atheist.

From single-cell organisms to humans, he found life in all its forms to be "extraordinary" and "miraculous"—and he often spoke of his awe at such vastly complicated creations as people, with their immune systems and brains and all of the other organs and tissues that, when they worked, were marvels indeed.

"The complexity of life and evolution is mind-boggling," he said. "When you think of how many chances there are for things to go wrong, it is so beautiful. *The Wisdom of the Body*—that was a book written by Walter Cannon. He talked about physiology: how everything functions right together. It's just so true. Every time you move your arm like that, your motor neurons stimulate the muscle contractions on this side; you have an equal number with an inhibitory motion to stop it from moving, so that gives you balance. Otherwise when you went to move, you'd slam your face. All those nerves are functioning and we don't have to think about it."

So was this strictly the result of evolution, people simply being the most advanced of the life forms that all traced their ancestry to the first single-cell organism to appear in the primordial sea? Did the Big Bang theory fully explain the origins of the universe? Was there some Higher Power?

"Certainly I could never deny the possibility that there's something supernatural that I can't understand to explain all this, because I can't," he said. "And much greater minds than mine can't."

Sachs found meaning in a remark by Albert Einstein that hung on his study wall: "Strange is our situation here on earth. Each of us comes for a short visit, not knowing why, yet sometimes seeming to divine a purpose." Said Sachs, "If Einstein didn't know, then how can we expect to?"

There was more to the Einstein quote than the part on Sachs's wall, and it pertained to people's conduct: "There is one thing we do know definitively: That we are here for the sake of each other."

Sachs embraced that philosophy.

"I believe it," he said. "I live by it, or try to."

And not for the sake of some future reward—Sachs sometimes tried to believe in an afterlife, but the belief never stuck. "You have to live for this life, there's no question about that," he said. "And we have to do everything we can to make this place a better place before we leave."

On January 10, Sachs turned sixty-two.

He was not a gloomy man, but lately he had been contemplating the time he had left—and the work remaining in so many fields that had to be completed before disease and suffering would be put in their place. Had the thawing end of cryonics been solved, he occasionally remarked, he might be tempted to have himself frozen in hopes of returning to experience the wonders of a distant age. The only drawback, he once joked, was that he would be alone. "The people I care about think it's crazy and wouldn't go with me!" he said.

Cancer research had interested Sachs as a young doctor, but he had concluded that even the best minds probably would not find a cure in his lifetime. "I've always felt that I should work on the most important problems that are solvable," he said. "I think there are problems more important than transplantation, such as cancer, but I've never had the conviction that they are solvable."

Tolerance in conventional transplantation, he had decided, was solvable—and he had helped demonstrate its potential in the handful of people who now lived drug free after their transplants.

Xeno was also solvable, he had decided long ago.

"It all made sense to me and it still does but it's been a lot harder than I expected," he said. "I thought I'd have all the answers in a few years and would move on to regeneration, which was the other thing I wanted to do."

Years before the advent of stem cell technology, Sachs had hoped to find the way to program the human body to grow replacement parts. Nature, he observed, has already shown the way.

"You take an amphibian," he said, "and you cut off its arm and it forms a limb bud, and then it differentiates into bone and muscle and nerve and skin and the whole thing grows back. Why shouldn't we be able to do that? The genes are there—they've just become [nonfunctioning] 'pseudogenes,' and we just have to figure out how to turn them back on."

But after more than forty years in transplantation science, Sachs doubted he would ever get to regeneration.

"I've been very sobered by how long it takes to do something—and how limited one's lifespan is and how sorry I am that I can't live longer," he said. "I keep reading about [regeneration], though. And I see what's happening in the field." And maybe, with luck, he would help get xeno into the clinic and have time left to begin researching regeneration. "But I don't think I'll get to the answer in my lifetime—unless I figure out a way to be frozen or something like that," Sachs laughed.

"I'm sure it will happen. In the meantime, I'm working on what I think is going to save a lot of people's lives."

The Nuffield Council

Cheryl Snow went home sixteen days after her transplant, and by the end of the summer of 2003, she was back working at Home Depot. A biopsy on one of her many outpatient checkups at Mass General showed no signs of rejection. For a while, she had to wear a mask when she went outside—an insignificant price to pay to escape a room that visitors could not enter without first sterilizing their hands.

"Just to be out in the fresh air—it was unbelievable!" she said. For the first time in years, she had energy and strength.

But autumn came, and Snow felt sick.

She was running a fever and, on the day she was readmitted to Mass General, she was vomiting so badly she could not take her medications. Testing disclosed the presence of cytomegalovirus, a member of the herpesvirus group, in her body. Most healthy people infected with CMV have few, if any, symptoms, but for patients like Snow, whose immune systems are weakened by immunosuppressive drugs, CMV can be fatal. In Snow's case, the source of infection was unknown—but CMV, like many other pathogens, can be transmitted through transplanted organs. Screening does not always reveal their presence—and not every pathogen is screened for, a deficiency that has led to transplant recipients being infected with rare diseases. "They're never 100 percent sure," Sachs said. "It's not like it will be with the xeno, where we'll have all those pigs tested upfront."

Snow was discharged after a week. She went back to work and back to her new daily routine of taking fourteen different drugs, including cyclosporine and two others to help prevent rejection. Snow was experiencing two common side effects: she was gaining weight, and she was growing extra hair on her face and arms. "There's nothing I can do," she said.

But those complications are relatively mild compared to others that are possible in an immunosuppressive regimen. Foremost is increased susceptibility to infection: a case of the flu that might send an ordinary person to bed could send a transplant recipient to the grave. "Number two is cancer," Sachs said. "There are a variety of tumors that are increased in incidence in patients on immunosuppression."

Other side effects—some treatable, some not—include kidney damage, liver damage, high blood pressure, peptic ulcers, and excessive gum growth. Another issue is chronic rejection: even with a battery of drugs, about 5 percent of transplanted organs are rejected every year, leaving a recipient in need of another transplant.

Another side effect is more difficult to quantify—the psychological one, which for Snow manifested itself most strongly during Thanksgiving and Christmas. It concerned the dead person whose heart now beat in her chest.

"I was very sad during the holidays," she said not long after New Year's Day. "I am so glad they are over. I was crying a lot because I was thinking about the donor family. I thought they must be very upset and how could I enjoy myself knowing that they are missing a loved one? It was very hard for me. My husband says I have been talking in my sleep for the past week. I never have done that. He can't make out what I am saying. He seems to think I am either on the table having a biopsy or I am dreaming about my heart transplant. It has been every night. A couple of nights ago I woke up and heard voices. I am not kidding. They were coming from in the air over my husband while he was sleeping. I wasn't afraid of the voices."

The weeks passed and Snow continued to feel disturbed. She had learned through the hospital grapevine that the donor was a twenty-two-year-old man from the Boston area. For a while after her transplant, she had wanted to know no more.

Now she was thinking about contacting the young man's family. The rules established by the United Network for Organ Sharing, the national organization that coordinates transplants, require a recipient to attempt to contact a donor's family by communicating through the network. There was no assurance that a family would reply.

"I have had many emotional moments, especially during the holidays," she said. "All of a sudden, I will start thinking about the family and start crying. I don't know who they are, what they look like, or where they live. I don't know what the donor looked like and I am curious to know more about this person. His personality, likes and dislikes, and so on. Maybe he had one goal he wanted accomplished in his life and I could help the family with it, I don't know. I would definitely want to give something back to his family. Look what they have given me. My life is an emotional roller coaster. I blame it on the medication, but I know it is more than that."

In 1996 the London-based Nuffield Council on Bioethics, which studies the ethical implications of advances in medical research, concluded a review of xenotransplantation. Among many other topics, the council examined the psychological impact of transplanting a dead person's organ into someone's living body. The council could have been writing about Snow:

"Stresses occurring as a consequence of transplantation

include the general stresses of hospitalization and surgery. More specific stresses include coping with the fear of rejection of the transplant or with infection; with the intrusive nature of immunosuppression and follow-up treatment; and with a change in image of the body. Different levels of significance are attached to different transplants: tissue transplantation, for example, is seen as much less significant than organ transplantation.

"Heart transplantation is seen as most significant, since so much symbolic importance is attached to that organ: It is the seat of emotions (especially love), courage, enthusiasm and innermost thoughts. Transplant recipients report being affected by thoughts of organ donors and their families. For some, it is disturbing that they have inside them an organ from someone who has died."

So how might the recipient of a pig heart react?

The council noted that many people already have received pig parts: porcine valves, commonly used in valve-replacement surgery. Few patients reported any concerns with having the valves inside their bodies. But these are animal parts that have been treated to make them nonviable: they are not living tissue. Vital organs, of course, are.

"It is difficult to predict how people's views of their bodies, and of their identities, might be affected by xenotransplantation," the council wrote. "On the one hand, the use of animal organs might eliminate any disturbing implications associated with having a human organ. On the other hand, receiving an animal transplant might cause different stresses. The response is likely to reflect notions of what it is to be a person, to be human and to be an animal. These notions are not uniform for this or any other society, but vary according to social and cultural background."

The Nuffield Council also examined the more philosophical issue of what exactly a person living with an animal organ was. All human? Part animal? Something else entirely? This was not Greek mythology, with creatures such as the centaur, part man and part horse. This could be a large part of the fantastic future of real-life medical science.

"With xenotransplantation," the council wrote, "an additional boundary, that between human and animal, will become blurred. Whether, or in what ways, this is perceived as a problem will depend on how the human being–animal boundary is defined and the significance that is attached to it. If the essence of humanity is seen as a capacity to transcend the level of organic existence, then a person's sense of identity should not, in theory, be threatened by a transfer of organs across species boundaries. The idea of xenotransplantation may become troublesome if there is not thought to be a strict division between humanity and bodily existence. In this case, to receive an organ from an animal might be seen as a mixing of one's human essence with that of the animal, and therefore a dilution of one's humanity."

The Nuffield Council took its charge seriously, but others saw a dark humor in the issues raised.

The anti-xeno Campaign for Responsible Transplantation posted several cartoons on its Web site. One showed a patient in conversation with his doctor; the patient has grown a curly tail that he isn't aware of yet. "Side effect? What side effect?" the patient says. Another cartoon shows one pig saying to another, "So have you filled out your organ donor card yet?" A third cartoon features a pig talking to a man: "I can't believe you science guys want to use us pigs for human organ transplants! You can't violate my rights! It's immoral! It's wrong! It's obscene!" The man says, "Not to mention it's gross." Says the pig, "That too!"

Snow was solidly behind Sachs's grand ambition. "I would have a pig heart transplant if it was available," she said. "I wouldn't have been in the hospital for all those months waiting for a heart. What peace of mind that would bring to someone who was waiting and wondering if they would be fortunate to receive a heart.

"And if I had a pig organ, I wouldn't be concerned about the family of the pig and I would not be crying about the death of a pig. There would be no psychological dealings."

A Ghostly Environment

On the afternoon of February 9, 2004, Sachs gathered his scientists for an update on the few remaining double-knockout experiments.

The last pig-to-baboon transplants had been performed on the difficult day of November 17, and no new ones were scheduled for at least two months. With Novartis gone, Infigen had stopped producing double-knockout pigs. Only four animals remained, and Sachs wanted to use them for breeding, in the hope that nature would pick up where the cloners had left off. Sachs had no expertise in cloning, nor money now to hire someone who did.

The February 9 meeting brought mixed news.

Baboon 229, which had received a pig heart on November 17, had died the week before, apparently of rejection. But the last remaining heart baboon, B228, transplanted on November 6, was doing fine. Ninety-five days after the operation, a recent biopsy had disclosed, the heart remained healthy; of the eight transplants Cooper had performed using double-knockout hearts, B228 was now the second-longest survivor. Only B223, whose pig heart had lasted 110 days, had gone longer.

The discussion turned to B134, the last baboon to receive one of Kaz's thymo-kidneys.

What happened was an unpleasant surprise, given the beginning course. The operation on November 17 had proceeded smoothly, without a hint of hyperacute rejection. As the weeks passed, B134 became something of a celebrity. Kaz used a minimum of drugs as the baboon progressed toward xeno tolerance, and tests of the kidney's creatinine levels consistently revealed normal function, with no evidence of longer-term rejection. This was the sort of outcome that prompted visions of clinical trials.

The scientists were so pleased with B134 that just two

days earlier they had shot a video of the animal, which they showed at the meeting. The animal sat contentedly in its cage, eating a banana and smiling—or what passed for such. This was about as close to cute as a baboon could get.

But yesterday B134 was found dead in its cage.

"It's a great disappointment," Sachs said.

The animal was necropsied, and slides of various organs were projected on the conference room screen. One slide showed that the animal's lungs had swelled, but another revealed that the thymo-kidney was a pure, healthy pink. B134's native kidney had been tied off, and Kaz's composite organ had been supporting the baboon's life since the transplant—eighty-three days, a record.

"Kidney looks perfect," said Megan Sykes. "What a tragedy."

One of Kaz's assistants projected a slide of the baboon's heart. It showed a scar, suggesting that acute myocardial infarction had killed B134.

"We have to figure out why this heart attack occurred and what we're going to do about it," Sachs said.

He sounded and looked weary. A drug reaction could have caused the heart attack, he theorized, although he later speculated that a small clot that had formed inside a catheter might have broken loose.

Almost a year had passed since the excitement of Goldie.

After some two dozen experiments, only one baboon, B228, survived with a double-knockout organ.

*

Cooper attended the February 9 meeting, one of his last; he was in transition to Pittsburgh, where he had accepted the offer to head the Starzl xeno program.

Cooper had learned that the Pittsburgh scientists had performed the world's first double-knockout transplants: of pig skin and pancreatic islets into monkeys in late September 2002, some five months before Goldie. But the first whole-organ transplant, a kidney, occurred after Goldie: on May 19, 2003, again with a monkey recipient.

None of the experiments, Cooper learned, had achieved results like those in the Sachs lab. "They didn't get very long survival," Cooper said, "a couple of weeks or so." Measured like this, Sachs was far ahead.

But in Cooper's estimation, the Pittsburgh program had potential. In addition to a continuing supply of double-knockout pigs, the Starzl group had expertise in genetic engineering.

And it had money from Revivicor's investors, which included Highmark Health Ventures Investment Fund (affiliated with Highmark, the managed care provider) and Fujisawa Investments for Entrepreneurship—the sort of capitalists Greenstein had courted.

"We believe that through Revivicor's business infrastructure we can make important scientific strides," said University of Pittsburgh Medical Center president Jeffrey Romoff. "We are indeed positioned to develop technologies that will have a significant impact on people's lives."

Immerge's imperiled state influenced Cooper's decision

to leave Boston. "If they didn't stay in business, then I knew we'd be doing much less here," he said. "The University of Pittsburgh purchased that part of PPL and they are prepared to put some money into it in order to try to get this technology to work. They know it's going to be expensive and yet they reassured me that they are in it for at least the medium to long haul. That means we're at least assured of getting some funding for the next few years. I wasn't sure what would happen here."

Although the financial picture in Boston troubled Cooper, he praised the advances that had been made during his seven and a half years with Sachs.

"When I started," he said, "we were giving a lot of treatment, and it was a huge procedure. We were doing plasma pharesis and removal of antibodies, we were giving irradiation, we were giving big doses of drugs—all this sort of stuff. Now the immunosuppressive regimen is very benign and yet we're getting out to at least two to three months. And the baboons are much healthier, and they're less and less stressed. There's virtually no infection now and we're not getting the major forms of rejection. Everything has progressed very significantly, particularly in the last two to three years.

"I'm much more optimistic about xenotransplant becoming a clinical entity now than I was two or three years ago. I think if we can identify some more genes which should be put in or taken out—and if we can sophisticate the immunosuppressive regimen—we may well be getting to some several months, if not longer, and then we could begin to think about a clinical program."

Although he now answered to a new boss, Cooper did not consider himself to be in competition with his old one. "There's so few of us left that I think we'll collaborate as much as we can. There's just a handful of places still really interested and with the resources to do anything."

Sachs agreed. Cooper had informed him after Christmas of his decision to leave, and Sachs had wished him the best. He maintained that if Cooper could help advance xeno in Pittsburgh, then everyone would benefit.

"I'd like it to be here but I'm happy to help anybody anywhere," Sachs said. "I've always been that way."

Unlike some in his field, Sachs was not close-lipped or paranoid. The story was told of one xeno researcher who sometimes slept in his lab to guard his animals and the experimental results.

Visitors were welcome in Sachs's lab, and the inscriptions on many of the photographs he hung on the walls of the dozens of fellows he had trained spoke to his role as a teacher, not a keeper of secrets. Sachs prided himself on his openness. He still carried in his briefcase a note from a fellow who had left years before: "In the scientific community there is considerable arrogance, condescension and backstabbing to be found among many of the leaders in a field," the young doctor wrote. "I have heard countless conversations in which a particular scientist has bad-mouthed another for whatever reason; I have heard only universal praise of you as a kind and considerate person who as a great scientist nonetheless helps others when possible."

It did not mean, however, that Sachs lacked an ego. Nothing would thrill him more than being first to break through.

Sachs had never seen a Pittsburgh/PPL double-knockout pig, but he was convinced that his own pigs were better, since they had been cloned from his three-decade-old line. "I think our knockouts are far better than theirs for a variety of reasons—they're inbred, they're miniature, they're complete knockouts," said Sachs.

But he conceded that money could give the Starzl group an advantage: while he now had to breed a line of double-knockouts before resuming intensive research, Cooper and his new group of scientists had a steady supply ready for experiments now.

"They have the big advantage of [double-knockouts] being available," Sachs said. "They could catch up with me and pass me in terms of where they are just by the sheer fact that I have to wait until I get my breeding going now before I can do many more experiments."

Novartis was out of the picture, but Greenstein had enough cash on hand to keep Immerge going into the new year—albeit with a reduced staff and diminished hope that she would be able to save the company. By the end of March, the hope was all but gone.

A visitor to the company at that time entered a ghostly environment. Where once scientists and technicians had sat at lab benches absorbed in their work, the handful of remaining employees were packing equipment into boxes. Some would go into storage, against the possibility that Immerge was resurrected, and the rest was being sold or loaned to other scientists. Hawley, the researcher who had knocked out both copies of the sugar gene, a revolutionary advance, was among those packing up. Near him, an alarm beeped on a freezer that had once stored tissue samples: the freezer had been unplugged and emptied, but no one knew how to disable the battery-operated alarm.

"This is not how I want to be remembered," Hawley said.

Greenstein was in her office, finalizing accounting and inventory details and attending to a myriad of other paperwork. Absent a receptionist or secretary, she also was answering the phone. Despite the atmosphere of defeat, she managed a laugh. "It feels psychotic," she said. "I mean, we're shutting down on one hand and we're actually having some interesting conversations with people. We are continuing to discuss financing options with a few players, but no deal yet."

Come April 1, Immerge would consist of boxed equipment, a Web site, an enormous volume of data, and two consultants: Greenstein and her chief scientific officer, both of whom would continue the quest for investors. But the summer passed without anything but talk.

The plight of Immerge helped bring an end to Infigen, and it slowed research at the University of Minnesota's Diabetes Institute for Immunology and Transplantation, which had collaborated with the Massachusetts firm in research using pig pancreatic islets to cure diabetes.

Starzl's group continued with Cooper. Researchers at a few centers kept on too, notably at the Mayo Clinic College of Medicine in Minnesota. But the Mayo Clinic no longer had the support of Baxter, which, like Novartis, had soured on xenotransplantation. Baxter donated its xeno program to the Mayo Clinic and walked away from the field.

With Immerge dormant, the xeno universe had shrunk further.

The report on B228 at animal rounds on April 2 was that the double-knockout heart seemed to be hardening but it continued to beat. It had been transplanted into the baboon 148 days before.

"This is now the world's record, isn't it?" Sachs said.

Kaz confirmed that it was, by a single day; the previous record, he said, was 147 days, held by Dr. Christopher G.A. McGregor of Mayo Clinic.

"It is starting to fail but it still is very nice," Sachs said. "We'll see what happens now."

Sachs scheduled a biopsy for May 4 to get a definitive analysis, but by the end of April, the baboon's health was declining rapidly. Rather than lose the host animal, Sachs decided to remove the alien heart immediately. Surgeons took it out on May 3, 179 days after it was implanted. It had lasted half a year, longer than any xeno heart ever.

"It was still weakly beating. But, you know, six months—just that it's alive is a very big achievement," Sachs said. He saw no reason why further research wouldn't extend the record, moving them closer toward the day when they could begin clinical trials.

Since returning to Harvard in 1991, Sachs had tried many approaches in his pig-to-primate experiments. He had perfused monkey blood through pig livers, circulated baboon blood through a silicon matrix, used some of David White's transgenic pigs. "All those things helped but nothing has worked as well as these knockouts," Sachs said. They probably needed some further genetic engineering and the protocols might need to be further refined, but these pigs were the answer, Sachs was convinced.

"This is the time really to be doing it all," he said. "We should be expanding with the liver, the heart, lung, kidney. At this point, we should be raising large numbers of these pigs. We should be getting various groups working on it together and going in all these directions. This is the time for this thing to really be moving forward.

"Instead, we're cutting back—I'm narrowing down to the minimum I can do to prove that this will work. It's so frustrating—so exciting but frustrating. If it hadn't worked, I could very easily have walked away from it. There's lots of other things I have here that are working that I'm excited about—the [allo] tolerance and all that. I could spend [more of] my time on that instead of beating my head against the wall to get funding."

But the dream continued to cast its spell. Sachs was not giving up.

CHAPTER 6

A Perilous Existence

Informed Consent

Starting with Goldie, pig-to-baboon operations were a monthly exercise during most of 2003. But nearly half a year had passed since the last such operation on the morning of May 12, 2004, when technician Sheils went to fetch a pig from the special cage in the larger of Sachs's two pig rooms. It was a double-knockout—one of the first that Sachs had bred from a herd of single-knockouts, precursors to the Goldie line. With cloning not an option, at least for now, it symbolized the future for Sachs.

The animal had no name, just the number 16013 etched on a tag clipped to its ear—but it had elicited an unusual degree of affection from the animal care and operating room staffs. Perhaps it was because they hadn't seen a double-knockout since November. Perhaps it was because the animal had a crossed eye and elicited sympathy.

"Hey buddy," said Sheils, as she stroked the animal's head.

Another technician gave 16013 a sedative, and it drifted off. Soon it was on an operating room table, awaiting Kaz's scalpel.

A few minutes later, a shaved and groggy baboon was brought into the room and placed on the other table.

"Hey, big guy!" Sheils said. B138 was a male that weighed about twenty-two pounds, a small one. Its remaining fur was gray tinged with brown, and its teeth were bared. It was not as appealing as a lazy-eyed pig that liked having its head stroked, but Sheils was tender nonetheless.

The radio was on this morning, tuned to a classic rock station; the Captain and Tennille's "Love Will Keep Us Together" was playing. The song seemed an odd accompaniment to what was unfolding inside the room.

The baboon was settled onto the table, and a technician started the anesthesia.

"Ready, buddy?" Sheils said as she held the animal's hand.

The baboon went under quickly.

At 9:00 A.M. exactly, the time specified on the day's schedule, Kaz, recently promoted to associate professor of surgery at Harvard Medical School, walked into the operating room.

Kaz shot photographs of the pig and the baboon, then cleaned and sterilized the pig and began to open her up. This was Kaz's second operation on the animal: six weeks before, he had grafted a piece of its thymus onto one of its kidneys, creating the thymo-kidney he would put into B138 today.

Removing the thymo-kidney took little more than an hour. Kaz placed it in a bowl of ice water, trimmed it, and announced, "Ready for transplant."

He brought the organ to B138. Other surgeons had already excised one of the baboon's kidneys, and Kaz set about sewing his creation into place. Then he tied off the animal's native kidney. He worked without hesitation and took no break.

"He's fast," said a technician.

"Very fast!" Kaz agreed.

By noon, he was done. Pink Floyd's "Brain Damage" was playing on the radio. Kaz surveyed his transplant and pronounced it "nice." Then he asked a technician to photograph him and his assistant by the baboon.

"Smile!" the technician said.

"One more, please," Kaz said. He wanted a close-up of the organ. "Transplant kidney is producing urine," he said.

It was half past noon: Kaz had finished according to schedule.

Pig 16013 was not left to die. Kaz had taken only the thymo-kidney, and the animal could live with its remaining kidney, which the surgeon might use in another experiment. The animal was brought back to its cage. Assistant Shannon Moran, who had shepherded Goldie off to surgery more than a year before, climbed in with it.

Feeling the aftermath of anesthesia, the pig tried unsuccessfully to stand. Moran cradled it in her lap.

"I know—it's been a tough day," Moran said. "It's all right. I'm right here."

As summer passed and no investors signed on, Sachs grew increasingly disappointed with Novartis. He was dismayed that the $25 billion pharmaceutical firm had spent so much on Imutran, whose transgenic pigs had not proved to be the Holy Grail—and now they had pulled the rug on Immerge and on him, the scientist whose pigs had proved superior.

"This is the time when they should be doing what they did years ago," Sachs said. "They built colonies in several parts of the world to raise their pigs. They spent hundreds of millions of dollars." Things would have been so much easier if Novartis had not signed off.

Although he wasn't finding corporate support, Sachs's efforts to obtain other funding were paying off: he was winning new government grants and securing renewals on existing ones. It was enough to keep his breeding program going and to conduct a handful of xeno experiments, but not enough to bring him to the next milestone anytime soon, namely, FDA approval for clinical trials. The FDA had already drawn up guidelines for organs and tissues, for it didn't view xenotransplantation as a blue-sky concept. It assumed that at some point, some researcher would present a plan of whole-organ transplants that the agency would have to act on.

Several years in the making, the guidelines were contained in a forty-four-page document issued by the agency's Center for Biologics Evaluation and Research in April 2003. Among the chief concerns were the risk of transmitting disease ("zoonotic infections"), the choice of donor animal, and the selection of patients. The FDA would require extensive testing of donor animals and recipient people before, during, and after xenotransplantation.

"Because of the potentially serious public health risks of possible zoonotic infections," the report stated, "you should limit xenotransplantation to patients with serious or life-threatening diseases.... You should limit candidates to those patients who have potential for clinical improvement with increased quality of life following the procedure. You should also consider the patient's ability to comply with public health measures as stated in the protocol, including long-term monitoring."

The possibility that a pig (or other donor animal) could transmit a disease before the disease was recognized particularly concerned the FDA. An unrecognized disease could spread widely before public health officials were onto it.

"Transmission of microbial agents from xenotransplantation products could lead to systemic disease (for example, infection or neoplasia).... In addition, transmission of infectious agents could result in outbreaks of zoonotic disease, silent transmission of latent viruses, or emergence of new strains of pathogens. Experience has shown that widespread horizontal or vertical transmission of new pathogens is possible before the pathogens are recognized (e.g., Human Immunodeficiency Virus)."

The FDA urged special attention to monitoring for PERV and required a contingency plan if the virus became active in a patient. Anyone sponsoring xenotransplantation trials would have to be prepared to suspend or stop them and work with the FDA and other officials in taking "additional actions if required for the safety of the recipient and intimate contacts and to address possible health risks." The FDA did not specify what constituted "additional actions," but in an extreme case, quarantine presumably was one of them. The FDA seemed to echo concerns about other biologic agents for which the Department of Homeland Security had responsibility.

The FDA was not the only government agency with responsibility for regulating xenotransplantation. Starting in the mid-1990s, when Salomon Brothers predicted a $6 billion xenotransplantation market, the U.S. Public Health Service spent years studying the implications of putting animal organs into people. The spread of disease and the informed consent of patients topped the list of PHS concerns. Reading sections of its thirty-page guideline, published in August 2001, it was hard not to think of a Michael Crichton novel.

Before anyone agreed to become the recipient of an animal organ, the PHS would require the transplant team—preferably in the presence of a patient advocate—to explain, verbally and in writing, the possible consequences. Patients would be told of the risk of known infectious agents, such as PERV, and of the risk of becoming infected with an unknown disease that might manifest itself only after a transplant. The PHS wanted patients not only to understand risks but to undergo "counseling regarding behavior modification" to ensure they did not transmit to others a disease that, theoretically, could show first symptoms months or years later.

And patients were obliged "to educate his/her close contacts" of this potential dark side. The transplant team should provide a "documented procedure" explaining how this education was to occur.

"Education of close contacts should address the uncertainty regarding the risks of xenogeneic infection, information about behaviors known to transmit infectious diseases from human to human (e.g., unprotected sex, breast-feeding, intravenous drug use with shared needles, and other activities that involve potential exchange of blood or other body fluids) and methods to minimize the risk of transmission. Recipients should educate their close contacts about the importance of reporting any significant unexplained illness through their health care provider to the research coordinator at the institutions where the xenotransplantation was performed."

Following transplant, recipients would be forbidden to donate blood, sperm, ova, or any other tissues or parts for human use. The prohibition would also apply to a recipient's sexual partners and even household members who might share a toothbrush or razor. Recipients were to have no contact with ordinary livestock of the source animal—regular farm pigs, in the case of the organs Sachs would supply. The PHS feared that since a patient was, in essence, part pig, he or she could be more susceptible to pig diseases. The agency did not address the issue of whether a person with a pig heart or liver could eat pork products—if, indeed, such an appetite could develop.

The PHS expected xeno recipients to consent to lifelong surveillance, including regular exams and the periodic archiving of samples of tissue and body fluids; if they moved, they would be required to inform health authorities, including the PHS and transplant sponsors such as Sachs, of their new address and telephone number. Even death would not bring privacy. Before accepting a xenotransplant, a patient would have to agree to an autopsy, even if the xenograft had been previously rejected or removed. "Advance discussion with the recipient and his/her family concerning the need to conduct an autopsy is also encouraged in order to ensure that the recipient's intent is known to all relevant parties."

The agency's final provision was to establish the Secretary's Advisory Committee on Xenotransplantation in the Office of Biotechnology Activities of the Department of Health and Human Services. This committee was charged with writing final policy and procedures.

What a long road since the freewheeling 1960s, when xenotransplanters were on their own. Now they and their patients, if and when they came, would be asked to carry a sacred trust. And Big Brother would be watching.

The government set a high standard, but some of what the recipient of an animal organ was expected to do was not dramatically different from the obligations on a conventional transplant patient—or, for that matter, a person infected with HIV. In pursuit of the better good, public health officials decided on checks and balances and then relied on the good intentions of people.

Sachs believed that with funding restored to the Immerge-era level, he could satisfy the government and might be able to go to clinical trials with whole organs in two to three years.

The question was, Which organ went first?

Based on Kaz's work, Sachs was leaning toward the kidney, but others were urging him to go for the heart or liver. Unlike patients with renal failure, who have dialysis to keep them alive, patients with failing hearts and livers have virtually no options beyond transplantation, of whatever kind. And those who pushed for hearts and livers wanted action soon. Until long-term organ survival was achieved, they envisioned an interim period in which a pig heart or liver could serve as a "bridge," keeping the patient alive until a human organ could be obtained.

"I think there's no question which organs have the most need," Sachs said. He meant hearts and livers. "But they also are the ones that probably would be most difficult. It will depend a little bit on how well our experiments go. Right now it would be the kidney because that's all I'm working on." And while he had transplanted pig livers into baboons, the pigs were not double-knockouts. He would have to start a liver program virtually from scratch.

Whichever organ went to the clinic first, Sachs or some company in partnership with him would need to open a specialized breeding facility before he could get FDA approval. Under the model that Sachs would likely use, only filtered air would enter the facility, and workers would shower and dress in clean clothing before going inside. Workers would be discouraged from having pets at home, would be required to avoid contact with other farm animals, and would be prohibited from duty if they were sick—even with a head cold.

The initial herd of donor pigs would be delivered by caesarian section, which would keep them away from pathogens that normally reside in the birth canal of a sow. Once delivered, the piglets would be transferred into an isolation chamber at the facility. Their food and water would be sterilized, and handlers would touch them only with rubber gloves inserted through the chamber walls. After a few weeks in isolation, the piglets, certified disease free through testing, would be moved to clean (but not sterile) quarters at the facility and would be allowed to breed. They and their offspring, the first donors, would continue to be monitored for disease—and one of about every fifty animals designated for transplantation would be a "sentinel animal." This pig would be euthanized and necropsied as a further, more intensive check against disease.

"We would have to get pigs that the FDA would say were adequate for us to put into patients," Sachs said. "And that would mean doing some caesarian sections and having clean animals—verifying that they don't have any pathogens, et cetera. We'd have to set up our screening program for PERV, which Novartis had but now is canceled. And then we'd have to convince the FDA we were ready to do it. You're talking about several years—at least two or three years. And that's if everything started working today.

"But that's a lot less than I used to say. I used to say ten years, so it's getting closer."

A Human Volunteer

Steve Hecht sat in a cavernous room in an architecturally bland building in Billerica, a suburb north of Boston. Like the twelve other people with him, he was connected to a hemodialysis machine by a tube plugged into his arm.

The staff at the center had tried to make the place look cheerful. They had taped colorful paper kites and flowers to the wall, but decorations went only so far: with its fluorescent ceiling lights and gray tile floors, the room had all the charm of a factory. And the audible pulsating of the machines, which stood as tall as a soft-drink dispenser, was hardly a soothing sound.

Three times a week, for about four hours each time, Hecht visited this place. A fifty-three-year-old chiropractor whose practice was in Concord, Massachusetts, Hecht suffered from end-stage renal disease, in which the kidneys stop functioning. The only cure is a transplant, and Hecht already had two; both had failed. "I've got three kidneys in me," he said, "but none of them work: my two original ones, and one transplant, the first transplant. The second one was taken out."

Some dialysis patients sleep during their sessions; some crochet or read their mail. Hecht passed his evenings reading and watching TV and sometimes wondering how Dr. Sachs's xeno work was progressing. He wanted to volunteer to be one of the first human recipients of a pig kidney.

"How are the baboons doing?" he said during a dialysis session in the spring of 2004. "I care about those baboons. That's my future."

His past was nothing anyone would envy.

Hecht was twenty-eight years old, writing screenplays and driving a cab in New York City, when his energy began to flag and his lower back started to hurt. He went to a chiropractor—a visit that prompted an interest in what he would make his profession—but he didn't improve. His condition worsened on a trip to Europe. "I was nauseous all the time," he said. "I had very low energy. My back was killing me. I was just really not doing well."

Back in America, he continued to deteriorate, with a cough he couldn't shake and a case of thrush. He couldn't sleep. His urine began to come out clear and then it stopped coming out altogether.

"That's when I went to the emergency room," he said. "If I had gone twenty-four hours later, I would have been dead because when you stop urinating all of the stuff that comes out of your tissues and your blood in your urine is not coming out. And the thing that will kill you is potassium in your blood—it will stop your heart."

The doctors diagnosed end-stage renal disease, the result of an autoimmune disorder of unknown cause in which Hecht's own immune system attacked and destroyed his kidneys. Hecht was placed on hemodialysis, which cleanses the blood. It was the autumn of 1979. Kidney transplantation was still an evolving discipline, with high mortality rates and imperfect immunosuppressive drugs. But no one offered Hecht the option.

Hecht tired of his visits to a dialysis center three times a week and he switched to peritoneal dialysis, in which a catheter is inserted into the abdomen and the peritoneal cavity is flushed clean several times a day. A patient can perform the procedure at home or at work.

"Your peritoneal membrane, which is a big membrane in your abdomen, acts as a filter," Hecht said. "The clear fluid absorbs all the toxins through the peritoneal membrane. Four times a day you drain that fluid out into another bag and replace it with fresh fluid. The risk is, you have a catheter that is open to the air when you do the transfer and you can get bacteria into your abdomen. And if you do that, you have peritonitis, which is a very, very dangerous condition. And that happened to me twice. First time, I almost died. The second time, they caught it in time."

Hecht went back on hemodialysis—and then he heard about cyclosporine, the new immunosuppressive drug that was being used experimentally at a few medical centers, including the University of Pittsburgh, where Starzl was the big name. Hecht called and was placed on the waiting list. It was late 1982.

"Five months later, I got the phone call: they had a kidney. I hopped on the plane from Portland, Oregon, to Pittsburgh and I got the transplant. I almost lost that transplant—I had very severe rejection, but it held on. And that kidney lasted five and a half years, until the end of 1988."

Before the kidney failed completely, Hecht's doctor bent the rules. "What he was supposed to do," Hecht said, "was wait for my kidney to fail, go back on dialysis, then put me on the list. But he put me on the list before it completely failed and I got called at five weeks for a second transplant."

The second kidney lasted eleven years. When it failed in late 1999, Hecht had no choice but to return to dialysis.

"After you reject two transplants," he said, "your immune system goes crazy. It goes into overdrive. It just starts pumping out antibodies against foreign bodies—organs and tissues. That's called being highly sensitized. So I am a highly sensitized transplant candidate and that means it's virtually impossible to find a match for me. I will reject anything but a perfect match."

People can survive for years on dialysis, but it is a burdensome and perilous existence. A patient can never be more than two or three days from a dialysis machine. A patient must eat a restricted diet and limit fluid intake. This is not someone who can have three cups of coffee in the morning and a six-pack over the weekend.

"I don't urinate at all," Hecht said, "so everything I drink stays in my body." In one dialysis session, the machine can remove as much as eleven pounds of fluid from his body, an exhausting experience: "When you walk out of here, you feel like a dishrag that's been wrung out." Pleasure comes in small doses: two or three times a year, when he is "really thirsty," he has half a bottle of beer. "But, you know, you down a beer and you've got to take it all out. You can only drink so much before you put yourself in real hot water."

Even the most conscientious patient is at risk of complications: a dialysis machine is at best an imperfect imitation of the living organ. Hecht suffers from insomnia, osteoporosis, and hyperthyroidism, all common side effects of dialysis. Other patients experience bleeding, clotting, infection, or anemia. Potassium levels can rise, endangering survival.

For some, it becomes too much.

"I've known patients who've killed themselves by not

coming here," Hecht said. "They didn't want to come and they didn't. And they died."

It was not, Hecht had read, an unpleasant end: it sounded like death by an opiate overdose. "It's not a bad way to die," he said. "You just kind of get more and more spaced out. As your blood gets more and more dirty, so to speak, you just start to lose consciousness. You can breathe; they don't need to keep you on any breathing apparatus or anything like that. You basically just fade into a coma."

Hecht learned of Sachs's research in 2003, when he read an article about xenotransplantation on the Internet. This was about the time that the first double-knockout experiments were under way. Cooper had been profiled in the piece, and Hecht telephoned him to say he was interested in volunteering. Cooper did not give details of the experiments, but he estimated that if all went well, clinical trials could begin in as soon as three years.

Hecht said that when they were ready to go, he'd like to be in the initial group. He would even consider being the pioneer.

"If the situation were somebody had to be first," Hecht said, "and they thought I was a good candidate and they said, 'Well, we don't really have any data to show that there's going to be a serious problem, there's nothing we know of'—I could very well do it." The possibility had appeal on another level. "There's something attractive to me about being a medical pioneer. I tend to be fairly fearless when it comes to that sort of thing. I was one of the first people on cyclosporine back in 1983."

And unlike a liver or heart recipient, Hecht would have dialysis to fall back on if a pig kidney failed.

He hoped, of course, for success—and he tried not to dwell on how long it could be before clinical trials began.

On June 27, B138 died of gastrointestinal bleeding, a complication unrelated to its thymo-kidney. The summer passed without the arrival of a white knight or the performance of another knockout experiment. Sachs kept his fellows busy concentrating on non-xeno research, especially tolerance, where he could legitimately stake a claim as pioneer. He had plenty of animals for that research: the herd he had started inbreeding three decades before now numbered some 450 pigs. Charles River Laboratories had housed Sachs's pigs, but a recent grant enabled them to be moved to a facility at Tufts University School of Veterinary Medicine.

On a Monday in late August, Sachs met with Kaz, veterinarian Mike Duggan, and Scott Arn, who tracked the breeding programs. The topic was the small number of double-knockout pigs that had been produced from the breeding program. Sachs looked weary. Kaz was unusually subdued.

Arn handed out a sheet of paper that displayed the status of forty knockout pigs. Starting in the late summer of 2002, Sachs had begun mating single-knockouts in the hope of producing double-knockouts. This was a hedge, a wise one, as it now happened: if cloning double-knockouts did not succeed, inbreeding single-knockouts would eventually deliver double-knockouts that he could use in experiments—and also establish a natural line. This was basic Mendellian genetics.

Arn's sheet showed only ten living double-knockouts from the breeding program. None had yet proved fertile. A double-knockout line had yet to be established.

"My overall, number one priority has to be not to lose the knockouts," Sachs said. "At this point we have to take the route of postponing experiments and save the line."

That pleased Duggan and Arn, who at this point did not want to use a double-knockout for anything other than breeding. Kaz, however, had conducted only one thymokidney experiment all year. He was anxious for more.

But Kaz did not put up a fuss; he simply asked when he might expect to have a pig.

Sachs looked at the sheet and his eye settled on animal 16188, a female double-knockout born on July 18. He gave Kaz permission to use it. Kaz would create two thymokidneys inside the animal on September 17 and transplant them into baboons in November.

"It's a risk, but we're going to do it," Sachs said.

The meeting ended, and Sachs returned to his corner office, with its magnificent view of Boston Harbor. The sun sparkled off the water.

"That was depressing," he said. "But the last thing I want to do is lose the animals."

September promised to be busy on the grant-writing

front, and Sachs still hoped for corporate support; recently a company had approached Mass General Hospital administrators with an interest in xeno. So far, it was only interest.

Sachs was not beaten—discouraged, but not beaten.

The xeno puzzle, he insisted, would be solved—if not by him, then by someone else.

He believed that with the same conviction the young polio victim had that he would walk again.

"I can't believe we won't get there," the scientist said. "I just hope it doesn't take longer than I've got to put into it."

Epilogue

On November 18, 2004, Kaz transplanted the two thymo-kidneys he had created in September into two baboons. One of them, B142, developed a clot in the renal vein and had to be euthanized. The other, B144, fared better, surviving an early rejection crisis and living in good health for more than two months, when a gastric ulcer caused liver damage that also required the animal to be euthanized. Sachs and Kaz found strong evidence, however, that they had achieved tolerance in B144. "A second (double-knockout) kidney transplant would have confirmed this hypothesis," Sachs said, "but we were not able to perform this operation in time." He did not have another donor pig to use.

The new year dawned and no xenotransplants were scheduled, but Sachs continued his pursuit of funding. He remained optimistic about having one of his major NIH grants renewed—and he was pleased when the Massachusetts General Hospital's Development Office joined the cause. The office arranged for a late-winter meeting at the Ritz Carlton hotel in Naples, Florida, of philanthropists who might be willing to help Sachs.

The new year also marked the first publication of Sachs's research with the double-knockout pigs. Two papers, on Cooper's heart experiments and Kaz's kidney work, were published in the journal *Nature Medicine*. Cooper reported that hearts transplanted from double-knockout pigs lasted longer than from any other kind of pig. Kaz and Sachs called the results with kidneys and thymus "very encouraging." They concluded: "At autoposy, the grafts appeared normal on gross inspection . . . suggesting that additional small changes to the treatment regimen may permit the kind of prolonged survival required for eventual clinical applications."

But they were still in the lab, not the clinic. They still needed money.

Acknowledgments

My sincere thanks to everyone who appears in the main body of this book. Each and every one welcomed me from the start, and their goodwill continued as the months turned to years and I was still there.

My highest gratitude belongs to David Sachs, who opened the door to me, and to David Cooper, who brought me to Sachs's door. None of us knew how events would unfold when I began—and it is to David's credit that when the story took some discouraging turns, he allowed me to stay. Thank you, David, you are a man of your word. Thanks also to David's wife, Kristina; his sister, Judy Shulman; and his longtime colleague and friend Steve Rosenberg.

Numerous people who do not appear in the body of the book helped behind the scenes, including these current and former members of Sachs's Transplantation Biology Research Center: Christene A. Huang, Adrienne B. Sisco, Annette Sugrue, Dr. James S. Allan, Dr. Douglas R. Johnston, Maria

Doherty, Dr. Robert Cina, Patricia Kiszkiss, Dr. Stuart Houser, Erica Altman, and RoseMary Koch.

At Mass General Hospital: special thanks to chief public affairs officer Peggy Slasman and to Michelle E. Marcella of her staff. Also, Dr. Thomas Spitzer. Thanks to Annie Moore and Anne Paschke of the United Network for Organ Sharing. And thanks to Nicole Bubendorff of Novartis.

I'd also like to acknowledge several people who helped me toward a better understanding of regenerative medicine: Dr. Joseph P. Vacanti, Dr. Anthony Atala, Dr. Wing Cheung, Dr. Michael Y. Shin, Robert Langer, Michael West, Dr. Peter J. Quesenberry, Brian Orrick, Jeff Borenstein, Mohammad R. Kaazempur-Mofrad, and the late Christopher Reeve, who granted me two interviews.

Thanks to Dr. Jerry Lemler, Hugh Hixon, Robert A. Freitas Jr., Peter S. DiStefano, Roy Walford, Karla Steen, Peter Voss, Doug Dollemore, and Barbara Gummere. At Brown University: Terrie Fox Wetle, Michael Lysaght, Dr. Patrick Sullivan, Scott Turner, John Sedivy, and Dr. Lee Edstrom.

My agent, Kay McCauley, was my biggest booster, as always.

At PublicAffairs, special gratitude to my editor, Lisa Kaufman, who stuck with me during the evolution of this book from a look at the Fountain of Youth to a narrative about xeno. Thanks also to Peter Osnos, Gene Taft, Chrisona Schmidt, Nina D'Amario, and Robert Kimzey. And thanks to Paul Golob, who left PA just as the book was getting off the ground.

At the *Providence Journal*, I'd like to thank publisher Howard Sutton; executive editor Joel P. Rawson; Managing Editor for

New Media Thomas E. Heslin; reporters Andrea Stape, Mike Corkery, and Jennifer Jordan, all of whom provided valuable input; head librarian Linda Henderson and her assistant librarian, Christina Siwy; and photographer Connie Grosch.

And finally, thanks to my family: my daughter Rachel M. Miller, who shares my enhusiasm for medical science; my other daughter, Katherine L. Miller, who helped with some of the research; my son, G. Calvin Miller, who was so patient during the hours I was hidden away in my study; and my wife, Alexis, who critiqued an early draft.

Notes

I observed all of the operations and all of the scenes involving Sachs, his scientists, and their work that are recounted in this book.

Chapter 1

Nearly 100 million pigs were slaughtered for food in the year 2001, according to the National Pork Producers Council, www.nppc.org.

Janet McCourt's story and quote were contained in "Making Medical Miracles Possible," *MGH2002: A Magazine for the Massachusetts General Hospital Community,* Boston, 2002.

Lanza's quote on xenotransplantation came from an interview on February 5, 2003, one of a series of interviews at his office and in his home.

Starzl was quoted in "As Cross-species Transplants Move Ahead, Some Scientists Call for Caution, Restraint," *Scientist,* August 21, 1995.

Caplan was quoted in "Researchers Try to Expand Use of Brain-dead to Help Give Life," *Pittsburgh Post-Gazette,* January 19, 2003.

Hattler's experiment was described in "Medical Research at Edge of Death Presents Quandry," *Baltimore Sun,* February 2, 2003.

Chapter 2

Statistics on dialysis patients are available from the National Institute of Diabetes and Digestive and Kidney Diseases, National Institutes of Health National Kidney and Urologic Diseases Information Clearinghouse,
http://kidney.niddk.nih.gov/kudiseases/pubs/kustats/index.htm.

Scheper-Hughes's editorial, "A Beastly Trade in 'Parts': The Organ Market Is Dehumanizing the World's Poor," *Los Angeles Times,* July 29, 2003.

Da Silva's story was recounted in a *New York Times* piece that was reprinted as "Organ Traders Thrive on Suffering Sale of Kidney: A Saga of Poverty and Desperation," in *South Florida Sun-Sentinel,* May 23, 2004.

Scheper-Hughes's major study of the organ trade, "The Global Traffic in Human Organs," was published in *Current Anthropology,* April 2000.

Parmly spoke on June 27, 2001, before the Subcommittee on International Operations and Human Rights, House International relations Committee. A transcript is posted on the Department of State Web site, http://www.state.gov.

Dr. Wang Guoqi's testimony before the committee was published on June 28, 2001, by the Associated Press.

Scheper-Hughes's final quote of the section came in an e-mail to the author, September 11, 2004.

PPL's press releases were all posted on the company's Web site, http://www.ppl-therapeutics.com/news/news_1.html.

PPL eventually published its cloning results in a peer-reviewed journal: "Cloned Pigs Produced by Nuclear Transfer from Adult Somatic Cells," *Nature,* September 7, 2000.

Immerge BioTherapeutics's press releases were posted on the firm's Web site, http://www.immergebt.com/press_room/2003_07_14.php.

Hawley's explanation of knocking out the sugar was related in an interview on April 1, 2004, and in an e-mail received on October 1, 2004.

Cooper's quote appeared in "Organ Transplants Get Lift," *Boston Globe,* January 12, 2003.

The pig cleanliness quote is from the National Pork Producers Council Web Site, http://www.nppc.org/resources/facts.html. The press release on shortening the wait for a transplant appeared on the same site.

The story of Cheryl Snow was based on interviews and observations by the author, interviews with her doctors, and a review of her medical record, which she agreed to let me read.

Cooper's background was drawn from his CV, several of his scientific papers, and interviews with him.

CHAPTER 3

Sachs's background was drawn from his CV, many of his scientific papers, his childhood scrapbook, and interviews with his principal scientists, his wife, his sister, and colleague Dr. Steven A. Rosenberg of the National Cancer Institute.

My brief portrait of Louis F. Fieser was drawn from http://moderntimes.vcdh.virginia.edu/PVCC/mbase/docs/napalm.html , which based its biography of the chemist on stories that ran in *Time,* January 5, 1968, and *Business Week,* February 10, 1969.

Dr. Keith Reemtsma's story is chronicled in Tony Stark, *Knife to the Heart: The Story of Transplant Surgery* (London: Macmillan, 1996).

The role of the thymus was reported in Kazuhiko Yamada, Pierre R. Gianello et al., "Role of the Thymus in Transplantation Tolerance in Miniature Swine. I. Requirement of the Thymus for Rapid and Stable Induction of Tolerance to Class I-mismatched Renal Allografts," *Journal of Experimental Medicine,* August 18, 1997.

The portrait of Mannheimer was based on "The Man Who Loved Monkeys: Hans Mannheimer Might Be Spinning in His Grave If He Knew What Was Happening to His Animals Today," *Ft. Lauderdale Sun-Sentinel,* March 24, 1996.

The aftermath of Hurricane Andrew was described in "Escaped Lab Monkeys Shot Out of Trees—But Most Lost Primates Have Been Recovered," *Seattle Times,* August 31, 1992; "In Andrew's Wake, A New Wild Kingdom: Monkeys, Cougars Still Running Loose Weeks After Storm," *Baltimore Sun,* September 20, 1992.

The profile of Kaz was created from interviews with him, his colleagues, and his CV and bibliography.

Budget data for the National Institutes of Health was found on the NIH Web site, http://www.nih.gov/about/almanac/appropriations/index.htm.

NIH grant data was from this NIH Web page: http://grants1.nih.gov/grants/award/trends/fund9303.htm.

I interviewed Dr. Boat by telephone on December 3, 2004.

Vacanti's observations on funding are taken from interviews on June 6 and November 20, 2003.

Quesenberry's comments on corporate research come from an interview on July 29, 2002.

Chapter 4

The best early history of xenotransplantation that I found was in David K.C. Cooper and Robert P. Lanza, *Xeno* (New York: Oxford, 2000).

A wealth of information about Brinkley was found in this biography: R. Alton Lee, *The Bizarre Careers of John R. Brinkley* (Lexington: University Press of Kentucky, 2002).

An account of Voronoff and his work was found in Tony Stark, *Knife to the Heart*. Additonal details, including Voronoff's quotes about the seventy-four-year-old man, were found in Lanza and Cooper, *Xeno*.

The most detailed account of the discovery and development of cyclosporine I have seen is at http://www.world-of-fungi.org/ Mostly_Medical/Harriet_Upton/Harriet_Upton.htm. My account is derived from it.

Novartis ranks cyclosporine as its third-best selling drug: http://www.novartis.com/products/en/bestsellers_list.shtml.

I read several accounts of the Baby Fae case. The "new vistas" quote was published in "Surgeon Hails Infant's Legacy," *Washington Post*, November 17, 1984. Bailey's philosophical ruminations were published in Tony Stark, *Knife to the Heart*.

The *Saturday Evening Post* first interviewed White for its September–October issue. The *Post*'s second interview of White was in the January 2000 issue.

Lebowitz spoke to the *Boston Globe* in its June 16, 1996, edition.

Sachs's editorial appeared in *Xenotransplantation*, August 1994.

The futuristic scenario of organ transplants was published in Robert P. Lanza, David K.C. Cooper, and William L. Chick, "Xenotransplantation: After Struggling for Decades with a Shortage of Donated Organs from Cadavers, Transplant Surgeons May Soon Have Another Source to Tap," *Scientific American*, July 1997.

I interviewed Paul Herrling by telephone on August 17, 2004, and he answered additional questions in several subsequent e-mails.

The BBC broadcast its report, *Welfare Groups Condemn Animal Transplants*, on October 13, 1998.

The Diaries of Despair is online at http://www.xenodiaries.org/index.html.

I interviewed PETA's Seidle by telephone on December 1, 2004.

The New York–based Campaign for Responsible Transplantation has a Web site at http://www.crt-online.org/index.html.

Singer's remarks were made in an e-mail received by me on December 3, 2004.

I attended the meeting of the Mass General Subcommittee on Research Animal Care that was held on December 18, 2002.

Bailey and Zola-Morgan spoke to Blum in "Viewpoint: Heroes in the Lab," in the Fall 1992 issue of *University of Georgia Research Magazine,* available online at http://www.ovpr.uga.edu/researchnews/92f/heroes.html.

I found results of the 1989 Gallup poll on use of animals in medical research in the Augist 22, 1989, issue of the (N.J.) *Record.* The 2004 results were published by the Gallup Organization on May 25, 2004.

The Vatican's Pontifical Academy for Life's "Prospects for Xeno-transplantation: Scientific Aspects and Ethical Considerations" is available online at http://www.vatican.va/roman_curia/pontifical_academies/acdlife/documents/rc_pa_acdlife_doc_2001 0926_xenotrapianti_en.html.

Herrling's quote on creating Immerge appeared in a September 26, 2000, press release.

Sachs spoke about the new company in "Merger Signals Shift in Xenotransplantation Research," *Nature Medicine,* November 2000.

Herrling's comments about islets were contained in an Immerge press release of June 3, 2003, as were Greenstein's comments.

I reconstructed Cheryl Snow's surgery from her medical record.

CHAPTER 5

Some of Cheryl Snow's statements about her mood during the 2003 holidays were made in e-mails to the author.

The Nuffield Council's 1996 study, "Animal-to-Human Transplants: the Ethics of Xenotransplantation," is available online at http://www.nuffieldbioethics.org/go/ourwork/ xenotransplantations/introduction; http://www.fda.gov/cber/gdlns/clinxeno.htm.

The xeno cartoons can be found at http://www.crt-online.org/humor.html.

The dates and details of Pittsburgh's first xenotransplants were provided to me by Cooper.

Romoff's remarks on the founding of Revivicor appear in a press release issued by the university of Pittsburgh Medical Center on April 8, 2003, and which moved on PRNewswire the next day.

CHAPTER 6

The FDA guidelines are available at http://www.fda.gov/cber/gdlns/clinxeno.htm.

The Public Health Service guidelines are available at http://www.fda.gov/cber/gdlns/xenophs0101.htm.

Index

Advanced Cell Technology, 13
Aging, ix
AIDS, 66, 87
    transplantation and, xiii, 129
Allotransplantation, 9, 74, 75, 113
    cyclosporine role in, 119
    Sachs and, 74, 75
    Snow and, 90–92, 152–153
AMA. *See* American Medical
    Association
American Medical Association
    (AMA), 115
American Veterinary Medical
    Association (AVMA)
    animal facilities accreditation
        and, 133
    euthanasia and, 137
Amnesty International, 35
Animal facilities accreditation,
    133
Animal rights activism
    animal suffering and, 129,
        131–133, 136

British Union for the Aboli-
    tion of Vivisection and, 130
Campaign for Responsible
    Transplantation and,
    140–141, 176–177
Compassion in World Farming
    and, 130
Diaries of Despair and, 131
disease and, 129
ethics and, 23–24, 77
Herrling and, 129
PETA and, 138–140
Uncaged Campaigns and, 131
xenotransplantation and, xiv,
    5, 24, 140–142
Arn, Scott, 204–205

Baboons, 16–17, 27–28
    antibody theory, 163
    baboon donors and, 88–89
    desirability of, 75–76
    diet of, 89

Baboons (*cont.*)
  diseases and, 76, 87
  disposition/handling of,
    27–28
  litter size of, 76
  Loma Linda University Med-
    ical Center and, 121–123
  pig heart recipients and,
    37–39, 52, 62–64, 160–163,
    178–179, 185–187
  pig thymo-kidney recipients
    and, 156–157, 189–190,
    207–208
  pig thymus recipients and,
    56–58, 156–157
  transplant survival in, 178–179,
    185–186
Baby Fae
  animal rights activism and,
    122–123
  ethics of, 122–123
  Knife to the Heart and, 123
  Loma Linda University Med-
    ical Center and, 122
  media coverage of, 122, 123
  xenotransplantation for, x–xi,
    121–124
Bailey, Leonard L.
  Baby Fae and, 122–123
  Loma Linda University Med-
    ical Center and, 122
Barnard, Christiaan N.
  Cooper and, 60–61
  Groote Schuur Hospital and,
    60
  heart transplants and, xii, 119
Baxter, 48, 185

BioTransplant
  Immerge and, 44–45, 145
  Lebowitz on, 43, 79–80, 125
  Novartis funding of, 79–80
  Sachs on, 43, 79–80, 145
  Sandoz funding of, 124
Boat, Thomas F., 106
Bone marrow, ix–x, 10, 19, 79, 81
  induced tolerance and, 11
  Sachs and, 79
Brigham and Women's Hospital,
    118
Brinkley, John R., 120
  AMA and, 115
  condemned prisoners and, 115
  Eclectic Medical School and,
    113
  goat glands and, 114–115
  placebo effect and, 116
  xenotransplantation and,
    113–118
British Union for the Abolition
  of Vivisection, 130

Campaign for Responsible Trans-
    plantation, 140–141,
    176–177
Caplan, Arthur, 22–23
Cardiomyopathy, 52–56
Center for Bioethics, 22–23
Center for Biologics Evaluation
    and Research, 192–194
Chimpanzees
  heart transplantation in, 121
  kidney transplantation in,
    73–74, 121

Cincinnati Children's Research Foundation, 106
Clinical trials, 196
    FDA requirements of, 191–193
    Imutran and, 128
Cloning
    Infigen and, 48, 49, 80, 84
    knockout pigs and, xv, 50–51, 80, 84
    PPL Therapeutics and, 40–49
CMV. *See* Cytomegalovirus
Colman, Alan, 42
Committee for Oversight of Research Involving the Dead, 22
Compassion in World Farming, 130
Cooley, Denton, 23
    on rejection, 121
    on xenotransplantation, 121
Cooper, David K.C.
    baboon antibody theory and, 163
    biography of, 59–62
    Goldie and, 39, 51–52, 56–58, 62–64
    Guy's Hospital and, 60
    on Kawaki, 28, 39, 56, 57, 62
    Nature Medicine and, 208
    on rejection, 79
    Starzl and, 180
    successes of, 181–182
    thymo-intestine and, 103
    xenotransplantation and, x–xii, 9, 14, 17, 59–60, 125–127, 157–160

Cosimi, A. Benedict, 29–30, 31, 36
Cryonics, x, 169–170
Cyclosporine
    allotransplantation role of, 119
    Hecht and, 201–202, 203
    immunological factors and, xi–xii, 119
    Sandoz and, xii, 119–120
    Snow and, 172
    Tolypocladium inflatum and, 119–120
Cytomegalovirus (CMV), 172

DeBakey, Michael, 23
    Foundation for Biomedical Research and, 137–138
Dialysis
    complications of, 202–203
    Hecht and, 32–33, 198–204
    hemodialysis and, 32–33, 198–204
    peritoneal dialysis and, 200–201
Diaries of Despair, 131
DiSalvo, Thomas G., 54
Disease risk, 130, 139
    animal rights activism and, 129
    in baboons, 76, 87
    FDA and, 192–193
    Hecht and, 199–200
    in miniature swine, 77
    prevention of, 126
    in primates, 77, 87–88
    White and, 125
    zoonotic disease and, 192–193

Dolly, the sheep, 19–20, 41, 42, 150
Donor policies, 139–140
Dor, Frank J. M. F., 37–38, 56
Double knockout pigs. *See* Knockout pigs
Dugan, Crystal, 25, 26, 51, 52
Duggan, Mike, 80, 204
  Goldie and, 11, 19, 39, 64
  transportation and, 15–16

Eclectic Medical School, 113
Einstein, Albert, 169
Ethics
  animal rights activism and, 23–24, 77
  Baby Fae and, 122–123
  Campaign for Responsible Transplantation and, 140–141, 176–177
  Caplan on, 22–23
  Catholic church on, 143–145
  Center for Bioethics on, 22–23
  Committee for Oversight of Research Involving the Dead on, 22
  Hattler on, 23
  knockout pigs and, 7–8, 55–56
  miniature swine and, 77
  Nuffield Council on Bioethics on, 174–177
  organ selling and, 32–34
  PETA on, 77
  transplantation and, xi, xii–xiii, 176–177

xenotransplantation and, xi, xii–xiii, 7–8, 24, 77, 176
Eugenics, 117
Euthanasia, 137, 207

FDA. *See* U.S. Food and Drug Administration
Feiser, Louis F.
  napalm and, 69–70
  Sachs on, 69–70, 167
Fishman, Jay A.
  disease prevention and, 126
  Xenotransplantation and, 126
Foundation for Biomedical Research, 137–138
Fujisawa Investments for Entrepreneurship, 180
Funding, 191
  by Baxter, 48, 185
  of BioTransplant, 43, 79–80, 124
  Boat and, 106
  collaboration in, 106–109
  Greenstein on, 183
  of Immerge BioTherapeutics, 145, 147–148, 155, 163, 183, 207–208
  MGH and, 106, 108, 160, 163
  NIH grants and, 104–105, 108, 109, 191, 205–206, 207–208
  by Novartis, 47, 79–80, 109–110, 145, 178, 191
  patent licensing and, 106
  Salomon Brothers study and, 20, 193
  by Sandoz, 124

savings in, 106
secrecy agreements and,
    106–107
of transplantation, xiv, 4–5,
    18–20
Vacanti on, 108–109
from venture capitalists,
    147–148, 155

alpha–1,3-galactosyltransferase
    gene, 42, 44, 45
Gene targeting, 44
Goat glands, 114–115
Goldie, the pig, 11, 13–19, 21, 49,
    50–51, 61, 80, 81, 179, 180
baboon recipients from, 28,
    37–39, 52, 56–58, 62–64, 81,
    83, 84, 85–96, 188–191
heart rejection in, 95, 96
Immerge BioTherapeutics and,
    49–51
kidney success in, 94
National Pork Producers
    Council and, 51
tolerance in, 95
transplantation day and,
    25–26, 37–40, 51–52, 56–58,
    62–64
transportation of, 15–16
Goodall, Jane, 28
Greenstein, Julia L.
funding and, 183
at Immerge BioTherapeutics,
    47, 49, 50, 145, 148, 150–151
Groote Schuur Hospital, 60
Guy's Hospital, 60

Hardy, James, 121
Harvard College, 68–69
Harvard Medical School
Sachs at, 69–70, 72, 165
transplantation and, xii, 4, 30
Yamada at, 99, 189
Hattler, Brack
artificial lung by, 23
ethics and, 23
Hawley, Robert J., 184
gene targeting and, 44
on knockout pigs, 39–40,
    43–44, 47–49
on spontaneous mutation,
    48
vectoring and, 44
Heart transplantation, xii, 19
baboon recipients of, 37–39,
    52, 62–64, 160–163, 178–179,
    185–187
by Barnard, xii, 119
in chimpanzees, 121
Knife to the Heart and, 123
by Lillehei, 22
Hecht, Steve
cyclosporine and, 201–202,
    203
disease symptoms of, 199–200
hemodialysis for, 32–33,
    198–204
peritoneal dialysis for,
    200–201
Herrling, Paul
animal rights activism and,
    129
on immunosuppression,
    145–146

Herrling, Paul (*cont.*)
   on Novartis, 128, 145–147,
     153–154
Highmark Health Ventures
   Investment Funding, 180
Hisashi, Yosuke, 26–27
Hospital for the Ruptured and
   Crippled, 5, 66
The Hot Zone, xiii
House Subcommittee on Inter-
   national Operations and
   Human Rights, 35
Hunter, John, xvi

Immerge BioTherapeutics, xiv,
   13–15, 18
   BioTransplant and, 44–45, 145
   business strategies of, 150–151
   cloning knockout pigs and,
     40, 44–45, 47–48, 50–51
   dormancy of, 184–185
   Goldie and, 49–51
   Greenstein and, 47, 49, 50,
     145, 150–151
   Hawley and, 39–40, 43–45,
     47–49
   immunosuppression and, 149
   National Swine Research and
     Resource Center and, 14
   Novartis funding of, 47,
     109–110, 145, 163, 178, 183,
     191, 207–208
   pancreatic islets and, 149
   PERV and, 146, 150–151
   Prather and, 40, 44–45, 47,
     48–50

venture capitalists and,
   147–148, 155
Immunological factors, 50
   cyclosporine and, xi–xii, 119
   immunosuppressant drugs
     and, 10
Immunosuppression, 154
   Herrling on, 145–146
   Immerge Biotherapeutics and,
     149
   knockout pigs and, 45, 78–79,
     99
   side-effects of, 10, 172–173
   tolerance to, 10–11
Imutran, 17, 18, 20, 191
   closure of, 145
   FDA and, 128, 129
   human trials and, 128
   Sandoz funding of, 124
   White and, 124, 127–128
Infigen, 185
   knockout pig cloning and, 48,
     49, 80, 84

James, Ron, 42, 46, 83

Kawaki, Kenji
   Cooper on, 28, 39, 56, 57, 62
   transplantation and, 162–163
Kidney transplantation, xi, 19, 81
   in chimpanzees, 73–74, 121
   Hecht and, 199, 201–202, 204
   kidney failure and, 32–33
   by Reemtsma, 73–74, 121
   by Starzl, 120–121

thymo-kidney and, 100,
156–157, 189–190, 207
Knife to the Heart, 123
Knockout pigs
breeding of, 197–198, 204–205
caesarian delivery of, 197, 198
cloning of, xv, 40–49, 50–51,
80, 84
double, 3–5, 6, 9, 13–15, 17–19,
21–22, 50–51, 156–157, 178,
188
ethics of, 7–8, 55–56
alpha–1,3-galactosyltrans-
ferase gene in, 42, 44, 45
Hawley on, 39–40, 43–44,
47–49
immunosuppression in, 45,
78–79, 99
rejection in, xv, 42, 45, 78–79,
80, 124
Sachs on, 3–5, 6, 8–20, 23–24,
154–155, 207–208
transportation of, xv, 3–4,
15–16
xenotransplantation with, xv,
7–8

Lanza, Robert P.
Advanced Cell Technology
and, 13
xenotransplantation and,
126–127
Lebowitz, Elliot
BioTransplant and, 43, 79–80,
125
Sachs on, 79–80

Lillehei, C. Walton
heart transplantation and,
22
Liver transplantation
by Starzl, 118–119, 121
by Vacanti, 108
Loma Linda University Medical
Center
baboons at, 121–123
Baby Fae and, 121
Bailey at, 122
Lung, artificial, 22, 23

Madsen, Joren C.
heart/thymus cotransplantation and, 134–137, 156–157
Snow and, 153
Subcommittee on Research
Animal Care and, 134–137
Major histocompatibility complex (MHC)
in immune system activation,
74–75
in tissue matching, 78
in transplantation, 74–75, 78
Mannheimer Primatological
Foundation
blood shortage and, 157
primate sources from, 85–87
safety/cleanliness at, 87–89
Massachusetts General Hospital
(MGH)
animal facilities accreditation
and, 133
collaborations at, 106, 108
DiSalvo at, 54

Massachusetts General Hospital
(MGH) (*cont.*)
funding at, 106, 108, 160,
207–208
Snow at, 52–56, 151
Subcommittee on Research
Animal Care and, 134–137
transplantation at, xii, 4, 10,
11
Mayo Clinic College of Medicine
Baxter funding of, 48, 185
McCourt, Janet, 10–11, 82–83
M.D. Anderson Cancer Center,
22
Medawar, Peter, 70
MGH. *See* Massachusetts General Hospital
MHC. *See* Major histocompatibility complex
Miniature swine, 50
desirability of, 76–78, 183
diseases in, 77
ethics of, 77
Miss O'Shea, 67
Moran, Shannon, 25, 190–191
Murray, Joseph E., 118

Nanobots, x
National Academy of Sciences
Institute of Medicine, 4
National Institutes of Health
(NIH)
animal center of, 76
grants from, 104–105, 108, 109,
191, 205–208
Sachs and, 74, 79

National Pork Producers Council, 51
National Swine Research and
Resource Center, 14, 43
Nature Medicine, 208
NIH. *See* National Institutes of
Health
Novartis, 18–19, 119. *See also* Sandoz
funding BioTransplant from,
79–80
funding Immerge from, 47,
109–110, 145, 163, 178, 183,
191, 207–208
Herrling at, 128, 145–147,
153–154
PERV and, 129, 198
Nuffield Council on Bioethics,
174–177

Organ donations
shortages of, 31–32
United Network for Organ
Sharing and, 31–32
Organ selling
black market in, 31–32
ethics of, 32–34
Organ trafficking
Amnesty International and, 35
condemned criminal organs
and, 34–36, 115
House Subcommittee on
International Operations
and Human Rights on, 35
World Medical Association on,
34–35

Organs, artificial, 12, 22
Organs Watch, 33, 36

Pancreatic islets, 149
Peer review, 40–41, 45–46
People for the Ethical Treatment
    of Animals (PETA)
    animal rights activism and,
        138–140
    donor policies and, 139–140
    on ethics, 77
PERV. *See* Porcine endogenous
    retrovirus
PETA. *See* People for the Ethical
    Treatment of Animals
PHS. *See* U.S. Public Health Ser-
    vice
Placebo effect, 116
Porcine endogenous retrovirus
    (PERV)
    benign nature of, 129
    FDA on, 193–194
    at Immerge BioTherapeutics,
        146, 150–151
    monitoring of, 193, 198
    Novartis on, 129, 198
    quarantine of, 193
    xenotransplantation and, 50,
        128–129, 146, 150–151
PPL Therapeutics, 19–20, 21, 23
    cloning pigs at, 40–49
    Colman at, 42
    James at, 42, 46
    peer review at, 40–41, 45–46
    purchase of, 181
    recAAT market and, 46–47

Starzl and, 43, 48–49
    xeno spinoff from, 46–47, 49
    xenotransplantation at, 40–49
Prather, Randall S.
    Goldie and, 49
    at Immerge BioTherapeutics,
        40, 44–45, 47, 48–50
    at National Swine Research
        and Resource Center, 43
Primates, xiii–xiv, 16–17
    desirability of, 75
    diseases in, 77, 87–88
    sources of, 85–87
Psychological impact, of trans-
    plantation, 174–177

Quarantine, 193
Queensberry, Peter, 109

Recombinant human alpha1-
    antitrypsin (recAAT)
    PPL Therapeutics and, 46–47
Reemtsma, Keith
    kidney transplantation by,
        73–74, 121
    rejection and, 121
    Sachs on, 72–74
    xenotransplantation by, 72–74,
        121
Regeneration, 170–171
Regenerative medicine, ix
Rejection, 112, 154. *See also*
    Tolerance
    Cooley on, 121
    Cooper on, 79

Rejection (*cont.*)
  Goldie and, 95, 96
  in humans, 10, 119
  in knockout pigs, xv, 42, 45,
    78–79, 80, 124
  in primates, 17, 57, 58, 178–179
  Reemtsma on, 121
  Ross on, 121
Rejuvenation by Grafting (1925),
  118
Revivicor
  funding by Fujisawa Invest-
    ments for Entrepreneur-
    ship, 180
  funding by Highmark Health
    Ventures, 180
  Starzl and, 151, 180
Ross, Donald, 121
Russell, Paul S., 30, 36–37

Sachs, David H.
  academics of, 68–70
  allotransplantation and, 74, 75
  on BioTransplant, 43, 79–80,
    145
  children of, 71–72, 164
  comeback plans of, 186–187
  Cosimi and, 29–30, 31, 36
  curiosity of, xvi, 67–68,
    167–169
  Dor and, 37–38, 56
  double knockout pigs and, 3–5,
    6, 8–20, 23–24, 154–155,
    207–208
  early biography of, 5–6, 65–68
  Einstein and, 169

  Feiser and, 69–70, 167
  funding of, 104–109, 155
  gardening and, 166–167
  Goldie and, 37, 38, 43–44,
    47–49, 57, 60, 64
  at Harvard College, 68–69
  at Harvard Medical School,
    69–70, 72, 165
  at Hospital for the Ruptured
    and Crippled, 5, 66
  influencers of, 167
  Lebowitz and, 79–80
  marriage of, 70–71
  at MGH, 28–31, 72, 74, 79, 93
  miniature swine and, 76–78
  Miss O'Shea and, 67
  National Cancer Institute and,
    68–70
  NIH and, 74, 79
  parents' deaths and, 92–93,
    165
  polio of, 65–68
  Reemtsma and, 72–74
  religion and, 167–168
  research progress of, 160–163,
    165, 196–197
  Russell and, 30
  secrecy of, 182
  staff hierarchy of, 8–9
  on transplantation immunol-
    ogy, xiii–xvi
  venture capitalists and, 155
  The Wisdom of the Body and,
    168
  in *Xenotransplantation,* 126
  on xenotransplantation,
    xiii–xvi, 72–74, 75

Salomon Brothers, 20, 193
San Francisco Chronicle (2001), 7
Sandoz. *See also* Novartis
  BioTransplant funding by, 124
  cyclosporine from, xii, 119–120
  Imutran funding by, 124
Scheper-Hughes, Nancy, 33, 36
Secretary's Advisory Committee
    on Xenotransplantation,
    195
Seidle, Troy
  on donor policies, 139–140
  PETA and, 138–140
  on public health, 139
Shiels, Meaghan, 89, 188, 189
  preop/operating room and, 25,
    26, 63
Shimizu, Akira, 162–163
Short-gut syndrome
  thymo-intestine and, 103
Snow, Cheryl
  allotransplantation for, 90–92,
    152–153
  cardiomyopathy in, 52–56
  CMV in, 172
  cyclosporine and, 172
  death room and, 91
  emotions of, 173–174
  false starts for, 90–92, 151
  immunospression side-effects
    in, 172–173
  at MGH, 52–56, 151
  United Network for Organ
    Sharing and, 174
  xenotransplantation and,
    55–56, 177
Spontaneous mutation, 48

Starzl, Thomas E., 20–22
  Cooper on, 180
  kidney transplantation by,
    120–121
  liver transplantation by,
    118–119, 121
  at PPL Therapeutics, 43, 48–49
  Revivicor and, 151, 180
  Thomas E. Starzl Transplanta-
    tion Institution and,
    20–22, 48–49
  in *Xenotransplantation* and,
    126
Stem cell research, ix–x, xii, 13
  adult, 12
  embryonic, 12
  funding of, 109
Swine. *See* Miniature swine

Thomas, E. Donnall, 118
Thomas E. Starzl Transplantation
    Institution, 20–22, 48–49
Thymo-intestine
  Cooper on, 103
  short-gut syndrome and, 103
  transplantation of, 102–104
  Yamada on, 102–104
Thymo-kidney
  survival of, 179, 204, 205
  transplantation of, 100, 154,
    189–190, 204, 205, 207
Thymus, 98
  cotransplantation of, 134–137,
    156–157
  T cell formation and, 82
  T cell-depletion in, 82, 100

Thymus (*cont.*)
  thymo-kidney and, 100, 154,
    189–190, 204, 205, 207
  tolerance in, 38–39, 81–82
  transplantation of, ix–x, 10, 19,
    81–82, 102–104
Tissue engineering, ix, xii, 12
Tissue matching, 78
Tolerance, 170
  bone marrow-induced, 11
  in Goldie, 95
  to immunosuppression, 10–11
  induced, 99
  McCourt and, 10–11, 82–83
  Sachs on, 70
  in thymus transplantation,
    38–39, 81–82
Tolypocladium inflatum, 119–120
Transfusions, blood, 111–112
Transgenic pigs, 17, 124
Transplantation
  AIDS in, xiii, 129
  Barnard on, xii, 119
  bone marrow, ix–x, 10, 19, 79,
    81
  capitalism and, xiv
  dialysis/kidney failure in,
    32–33
  early attempts at, 111–113
  ethics of, xi, xii–xiii, 176–177
  at Harvard Medical School, xii,
    4, 30
  of hearts, xii, 19
  heart/thymus cotransplanta-
    tion and, 134–137, 156–157
  immunological factors in,
    xi–xii
  of kidneys, xi, 19, 81

Medawar and, 70
  at MGH, xii, 4, 10, 11
  National Academy of Sciences
    Institute of Medicine and,
    4
  need for, 29–31
  pancreas-to-kidney, 83–84
  of testicles, xi, 114–115
  of thymo-intestine, 102–104
  of thymus, ix–x, 10, 19, 81–82,
    98, 102–104
  at Transplantation Biology
    Research Center, xii, 4
Transplantation Biology
    Research Center, xii, 4,
    61–62
Transplantation immunology,
    xiii–xvi

Uncaged Campaigns, 131
United Network for Organ Shar-
    ing, 31–32, 174
University of Pittsburgh, 181
University of Pittsburgh Medical
    Center, 20, 22
U.S. Food and Drug Administra-
    tion (FDA)
  Center for Biologics Evalua-
    tion and Research and,
    192–194
  clinical trials and, 191–193
  Imutran and, 128, 129
  on PERV, 193–194
  prohibitions of, 194–195
  regulations of, 192–193
  on zoonotic disease, 192–193
U.S. Public Health Service (PHS)

Advisory Committee on Xeno-
plantation and, 195
regulations of, 193–195
Secretary's Advisory Committee on Xenotransplantation
and, 195

Vacanti, Joseph P., 108–109
Vectoring, 44
Venture capital, 155. *See also*
Funding
Highmark Health Ventures
Investment Funding as, 180
Immerge BioTherapeutics and,
147–148, 155
Voronoff, Sergei, 120
on eugenics, 117
Rejuvenation by Grafting
(1925) by, 118
on xenotransplantation,
116–118

White blood cells, x
White, David J.G.
on diseases, 125
on Imutran, 124, 127–128
on transgenic pigs, 124
Winter, James, 57–58, 64, 157–160
The Wisdom of the Body (Cannon), 168
World Medical Association,
34–35

Xenotransplantation, 4–5, 7–8,
13, 20, 21

animal rights activism and,
xiv, 5, 24, 140–142
Baby Fae and, x–xi, 121–124
Brinkley on, 113–118
cellular signaling and, 12
Cooley on, 121
Cooper on, x–xii, 14, 17
ethics in, xi, xii–xiii, 7–8, 24,
77, 176
Lanza on, 126–127
PERV and, 50, 128–129, 146,
150–151
at PPL Therapeutics, 40–49
primate desirability in, 75–76
public health and, 12
Reemtsma on, 72–74, 121
regeneration v., 170–171
Ross on, 121
San Francisco Chronicle on, 7
Voronoff on, 116–118
Xenotransplantation, 7–8, 125–126

Yamada, Kazihiko
academics of, 97
early biography of, 96–97
Goldie and, 26–28, 37–39, 59
at Harvard Medical School, 99,
189
on heart/thymus cotransplantation, 156–157
on Hisashi selection, 26–27
in *Nature Medicine,* 208
surgical skills of, 97–98
thymo-intestine and, 102–104
thymo-kidney and, 100,
156–157, 189–190, 207
on xenotransplantation, 9, 19

G. WAYNE MILLER is a staff writer at *The Providence Journal*, where he has won numerous awards for his writing. He and colleagues at *The Providence Journal* were finalists for the 2004 Pulitzer Prize in Public Service. Visit Miller at www.gwaynemiller.com.

PublicAffairs is a publishing house founded in 1997. It is a tribute to the standards, values, and flair of three persons who have served as mentors to countless reporters, writers, editors, and book people of all kinds, including me.

I.F. STONE, proprietor of *I. F. Stone's Weekly*, combined a commitment to the First Amendment with entrepreneurial zeal and reporting skill and became one of the great independent journalists in American history. At the age of eighty, Izzy published *The Trial of Socrates*, which was a national bestseller. He wrote the book after he taught himself ancient Greek.

BENJAMIN C. BRADLEE was for nearly thirty years the charismatic editorial leader of *The Washington Post*. It was Ben who gave the *Post* the range and courage to pursue such historic issues as Watergate. He supported his reporters with a tenacity that made them fearless and it is no accident that so many became authors of influential, best-selling books.

ROBERT L. BERNSTEIN, the chief executive of Random House for more than a quarter century, guided one of the nation's premier publishing houses. Bob was personally responsible for many books of political dissent and argument that challenged tyranny around the globe. He is also the founder and longtime chair of Human Rights Watch, one of the most respected human rights organizations in the world.

For fifty years, the banner of Public Affairs Press was carried by its owner Morris B. Schnapper, who published Gandhi, Nasser, Toynbee, Truman and about 1,500 other authors. In 1983, Schnapper was described by *The Washington Post* as "a redoubtable gadfly." His legacy will endure in the books to come.

Peter Osnos, *Publisher*